Design of Structural Elements with Tropical Hardwoods

Abel O. Olorunnisola

Design of Structural Elements with Tropical Hardwoods

Springer

Abel O. Olorunnisola
Department of Wood Products Engineering
University of Ibadan
Ibadan, Oyo State
Nigeria

ISBN 978-3-319-65342-6 ISBN 978-3-319-65343-3 (eBook)
DOI 10.1007/978-3-319-65343-3

Library of Congress Control Number: 2017948214

Printed on acid-free paper

This Springer imprint is published by Springer Nature
The registered company is Springer International Publishing AG
The registered company address is: Gewerbestrasse 11, 6330 Cham, Switzerland

With love to my late mum, Esther Tanimowo Olorunnisola (1935–2005), a mother in a million

Preface

The primary objective of writing this book is to provide basic information on the design of structures with tropical woods. While this book is intended primarily for teaching university- and college-level courses in structural design, it is also suitable as a reference material for practitioners. The language of the text has therefore been tailored to suit the varied needs of these potential users. Although parts of the background material relate specifically to West and East Africa, the design principles apply to the whole of tropical Africa, Latin America and South Asia.

The most common wood products employed in structural applications in the tropics remain roundwood in form of logs, poles and posts; sawn wood (lumber); and to some extent glued laminated lumber (glulam). These wood products are, therefore, the central focus of this book. Newer wood products such as laminated veneer lumber (LVL), parallel strand lumber (PSL), laminated strand lumber (LSL) and cross-laminated timber (CLT) are only mentioned in passing.

This book is laced with ample illustrations including photographs of real-life wood structures and structural elements across Africa that make for interesting reading. It has numerous manual and *Excel* spreadsheet worked examples and review questions that can properly guide a first-time designer of wooden structural elements. A number of design problems were also solved using the FORTRAN programming language. This was considered useful for readers who are already familiar with that programming language and could serve as a guide for those who would prefer programming using other languages. Topics covered in the 13 chapters of this book include a brief introduction to this book, the anatomy and physical properties of tropical woods; a brief review of the mechanical properties of wood, timber seasoning and preservation, uses of wood and wood products in construction; basic theory of structures and structural load computations; design of wooden beams, solid and built-up wooden columns, wood connections and wooden trusses; as well as a brief introduction to the design of wooden bridges.

The writing of this book required consultation with numerous reference materials as indicated in the bibliography and critiques from several generations of undergraduate and postgraduate students in the Departments of Agricultural and Environmental Engineering as well as Wood Products Engineering at the

University of Ibadan, Nigeria, who took my classes in the design of wood structures and who had access to the lecture notes that were later converted into the manuscript for this book. The style of writing adopted for this book was positively influenced by students' comments and criticisms. The language became simpler and more lucid. I am sincerely grateful to all the authors whose works I consulted and my past and current students whose efforts made the writing much easier than it would otherwise have been.

Ibadan, Nigeria
March 2017

Prof. Abel O. Olorunnisola

Contents

About the Author

Prof. Abel O. Olorunnisola formerly of the Department of Agricultural & Environmental Engineering Faculty of Technology, University of Ibadan, Nigeria, currently heads the newly created Department of Wood Products Engineering of the same university. He holds a first degree in Agricultural Engineering and Master's and Ph.D. in Wood Products Engineering. He has about 25 years of experience in teaching undergraduate and postgraduate courses in Wood Products Engineering. He has been a visiting scholar to a number of universities and top research institutes in the USA, England, India and Brazil at various times. He served as the chairman of the Nigerian Society of Engineers, Ibadan Branch, between 2010 and 2011 and concurrently as the dean of the Postgraduate School, University of Ibadan, and the President of the Materials Science & Technology Society of Nigeria between 2010 and 2014.

Chapter 1
Introduction

Wood has, from antiquity, played a significant role in human economy. It has continued to serve man in diverse ways, even in the most advanced economies, in industries ranging from building construction to aerospace. It has become evident in recent times that the higher the level of economic development of a country, the greater her dependence on wood and wood products—both conventional and less readily recognizable items such as paper and other wood pulp products. The well-established forest industries in many timber-rich African countries include those involved in logging and timber extraction, sawmilling, manufacture of wood-based panel products (i.e. plywood and particleboard), furniture making, paper making, match making, wood preservation, carpentry and joinery, wood moulding and the manufacture of various wooden items such as tool handles, sport goods, weaving equipment, and wooden toys. The continental distribution of tropical forests is: 56% in Latin America, 23% in Asia and 18% in Africa. The forest zone of West Africa covers all of Liberia, Sierra Leone, most of Guinea, the southern halves of Cote d'Ivoire and Nigeria, and parts of Ghana, Togo and Guinea-Bissau. The East African coastal forests are comprised of the Northern and Southern Zanzibar-Inhambane coastal forest mosaics, characterized by tropical dry forests within a mosaic of savannas, grassland habitats and wetlands areas. The diversity in vegetation types across Africa, ranging from mangrove swamp along the southern coasts through freshwater swamp, to lowland rain forest and savanna, makes possible the growth of over 1000 tropical hardwood species, many of which grow to merchantable dimensions.

Wood is also one of the oldest and best-known structural materials and has been used for thousands of years. Despite the development of newer materials, it remains a primary and very widely used structural material given its excellent performance characteristics and affordability. It is extremely versatile and requires less energy to produce a usable end product than do other materials. Besides, with wood, the boundaries of design and experimentation may be pushed more easily than most other materials. It allows designers to introduce a special ambience and gracefulness to large buildings while fulfilling demanding structural requirements. The

A.O. Olorunnisola, *Design of Structural Elements with Tropical Hardwoods*,
DOI 10.1007/978-3-319-65343-3_1

development of manufactured wood products has, in addition, enabled wood to be used in many landmark large span structures.

The various uses of wood and wood products in construction works are typically classified into three, i.e. decorative, functional and structural uses. Many types of tropical woods have an especially decorative appearance as, for example, the *Khaya* and *Dalbergia spp*. They have a wide variety of colours ranging from white (*Terminalia superba*) to black (*Diospyros spp*). Also, quite a number of tropical hardwoods are extremely naturally durable, as, for example, *Lophira alata*. Logs, lumber and other wood products are traditionally put into structural use as beams, joists, girders, floor decks, columns, purlins, rafters, and trusses. Many other newer Wood products are commonly employed in structural applications across the globe. The functional uses of wood include wall partitions, ceiling, doors and door frames, windows and window frames, among others, while a major decorative use of wood is as fascia board. In many developing countries, wooden rafters are more common than wooden trusses. Some of the different forms of wood utilization are shown in Figs. 1.1 and 1.2, while Table 1.1 shows a list of tropical hardwoods utilized for different purposes in building construction in Nigeria and many other West African countries.

Across tropical Africa, Asia and Latin America wood, especially sawn wood of various sizes, remains a popular construction material in residential and non-residential buildings (office complexes classrooms, lecture theatres, meeting halls and worship centres). Some of the frequently traded species are shown in Table 1.2. The more durable species are typically used in outside construction, while the less durable ones that have not been treated are primarily used in interior construction. In a few countries, wood-framed buildings supported on either concrete pile (Fig. 1.3) or a sandcrete block (Fig. 1.4) foundations still exist. However, this is not

Fig. 1.1 Wooden rafter installation in a walkway

Fig. 1.2 Wooden truss installation in a workshop

Table 1.1 Tropical hardwoods utilized in building construction in Nigeria

Wood species	Sawnwood dimensions (mm)	Mode of utilization in building construction	Forms of utilization
Cordia millenni	25 × 300	Fascia board	Decorative
Afzelia africana, Ceiba pentandra, Cordia millenni	75 × 100	Wall plate	Structural
Afzelia africana, Terminalia ivorensis, Cordia millenni	50 × 150	Tie beam	Structural
Gossweiledendron balsamiferum	50 × 150	Rafter	Structural
Lovoa trichilioides, Terminalia ivorensis	50 × 75	Purlins	Structural
Tectona grandis, Triplochiton scleroxylon	50 × 50	Noggins	Functional
Afzelia africana, Khaya ivorensis, Cordia millenni, Nesogodonia papaverifera	50 × 250	Door and window frames	Functional
Triplochiton scleroxylon	25 × 300	Framework	Structural
Terminalia ivorensis, Triplochiton scleroxylon	50 × 75	Formwork	Structural

Source A market survey conducted by the author in Ibadan, Nigeria in 2014

a popular form of building construction, largely because of the cultural preference for concrete construction, the general but erroneous belief that wood construction constitutes a serious fire hazard, the challenge of termite infestation and the thermal discomfort associated with such buildings in humid conditions.

Table 1.2 Selected frequently traded timber species from Africa, Asia, Latin America

Botanical name	Natural durability	Resistance to impregnation	Drying rateb	Main usesa
Terminalia superba	Non-durable	Permeable	R	1; 8; 10
Mitragyna stipulosa	Non-durable	Permeable	R	7; 10
Afrormosia	Very durable	Extreme	RS	2, 6
Afzelia spp.		Resistant	FR	3; 4; 5; 6; 9
Gossweilerodendron balsamiferum	Very durable	Very resistant	FR	4
Triplochiton scleroxylon	Non-durable	Moderate	FR	1; 3; 7;10
Nauclea diderrichii	Very durable	Resistant	FR	2, 5
Terminalia ivorensis				7; 9; 10
Lophira alata	Very durable	Extreme	VS	2; 5; 6
Entandrophragma angolense	Moderate	Extreme	R	7, 10
Guarea spp.	Durable	Extreme		3; 9
Milicia excelsa	Very durable	Resistant	FR	1; 2; 3; 4; 5; 6; 9
Khaya ivorensis	Durable	Resistant	FR	1; 2; 3; 9; 10
Entandrophragma candollei	Moderate	Moderate	RS	7
Khaya senegalensis	Durable	Very resistant	FR	2; 7
Tieghemella heckelii	Very durable	Resistant	FR	1; 2; 7; 10
Mansonia altissima	Very durable	Very resistant	FR	6; 7; 8; 10
Lovoa trichilioides	Moderate	Resistant	FR	7; 10
Entandrophragma cylindricum	Non-durable	Resistant	FR	1; 2; 7; 9; 10
Pterygota macroparpa	Non-durable	Permeable	RS	1; 8; 10
Entandrophragma utile	Durable	Extreme	FR	2; 4; 9; 10

Sources Okigbo (1964), Market Survey by the author
[a]*Main uses*
1 = interior construction
2 = boat building
3 = packing, containers, cigar boxes
4 = exterior construction
5 = sleepers, gardening, agriculture
6 = flooring, parquetry
7 = living-room furniture
9 = windows
8 = bedroom, kitchen furniture
10 = doors
[b]*Drying Rate*
R—Rapid
FR—Fairly rapid
RS—Rather slow
S—Slow
VS—Very slow
Source Market survey by the author (2013)

Fig. 1.3 A wood-framed building on concrete pile foundation

Fig. 1.4 A wood-framed building on sandcrete block foundation

The design of structural elements in wood is similar to the process adopted for steel, concrete and other well-established construction materials. However, textbooks on the subject matter are not as commonly available as those published on the other aforementioned construction materials. To date, students, practising engineers, wood technologists, architects, builders and other interested readers in sub-Saharan Africa do not have access to textbooks on the design of structures with tropical hardwoods. This book is designed to meet part of the textbook needs of students and professionals. Although parts of the background material relate specifically to West and East Africa, this book's principles apply to the whole of tropical Africa, Latin America and South Asia because, while building traditions may vary, the available materials are similar.

Also, a number of software packages have been developed for structural design with wood, but these are largely unavailable in many African countries. However, spreadsheets have become one of the most popular forms of computer software, second only to word processors. Spreadsheet software allows the user to combine data, mathematical formulae, text and graphics together in a single workbook. For this reason, spreadsheets have become indispensable tools for engineering design. *Microsoft Excel* in particular has won acclaim for its ease of use and power. As *Excel* has expanded in power and ease of use, there has been increased interest in its use in the classroom. There are also many advantages to using *Excel* in an introductory structural design course. An important advantage is that students are more likely to be familiar and more comfortable working with data entered into it.

Since *Excel* is very common at colleges and universities across Africa where structural design packages are scarce, an instructor can teach a course without requiring students to purchase an additional structural design software package. This book introduces readers to the use of ***Microsoft Excel*** for structural design with wood. A few design examples and problems based on the *FORTRAN* programming language are also included. It has numerous *Excel* spread sheet worked examples and review questions that can properly guide a first-time designer of wooden structural elements. However, readers should note that *Excel* is not a structural design package, and there are limits to what it can do in replacing a full-featured design package. A brief review of some simple *Microsoft Excel* operations is provided in Appendix A.

References

Lucas EB (1983) Factors preventing wider commercialization of Nigerian trees. For Prod J 33 (5):64–68

Lucas EB, Olorunnisola AO (2002) Wood processing and utilization in Nigeria: the present situation and future prospects. In: Ajav EA, Raji AO, Ewemoje TA (eds) Agricultural Engineering in Nigeria: 30 years of University of Ibadan experience. Published by the Department of Agricultural Engineering, University of Ibadan, Nigeria, pp 98–109

Lucas E, Olorunnisola A, Adewole N (2006) Preliminary evaluation of guava (*Psidium guajava* L.) tree branches for truss fabrication in Nigeria. Agricu Eng Intern: The CIGR E-journal manuscript No. BC 05 010, 8:1–10

Okigbo L (1964) Some Nigerian woods, 2nd edn. Published by the Federal Ministry of Information, Lagos, Nigeria

Chapter 2
Anatomy and Physical Properties of Tropical Woods

2.1 Anatomical and Gross Features of Hardwoods

Wood and its derived products are used in diverse ways including building construction; manufacturing of furniture items, tools, weapons and aircraft parts; and combustion as fuel, among others. In spite of its versatility and exceptional qualities, a lot of prejudices, oftentimes based on gross exaggerations of the inherent disabilities of wood species in general, still persist. Sound knowledge of the characteristic qualities of wood as an engineering material is therefore essential to its structural utilization.

New wood is produced by the formation of new woody layers between the existing wood and the inner bark of trees, resulting in increase in diameter. Though wood usually forms the stem of a tree, it is not all plants that possess woody stems. It is also not all woody stems that produce timber suitable for use as industrial material. For a plant to qualify as a woody plant, it must be perennial; possess the vascular bundle, i.e. xylem (wood), phloem (inner bark) and cambium (Fig. 2.1); and exhibit secondary thickening otherwise known as growth in diameter.

As shown in Table 2.1, wood in general is composed primarily of carbon, hydrogen, oxygen and nitrogen. The elemental compositions of selected tropical hardwoods and the typical chemical composition of hardwoods in general are shown in Tables 2.2 and 2.3.

These basic chemical elements are incorporated into a number of organic compounds known as cellulose, a crystalline polymer derived from glucose (constituting the wood fibres which are strong in tension), hemicelluloses, lignin (a matrix which resists compression), extractives and ash forming materials.

Woody plants that produce merchantable timber are divided into two broad groups, i.e. ***softwoods*** and ***hardwoods***, based on differences in anatomy and chemical composition as shown in Table 2.3. For example, softwoods (also known

A.O. Olorunnisola, *Design of Structural Elements with Tropical Hardwoods*,
DOI 10.1007/978-3-319-65343-3_2

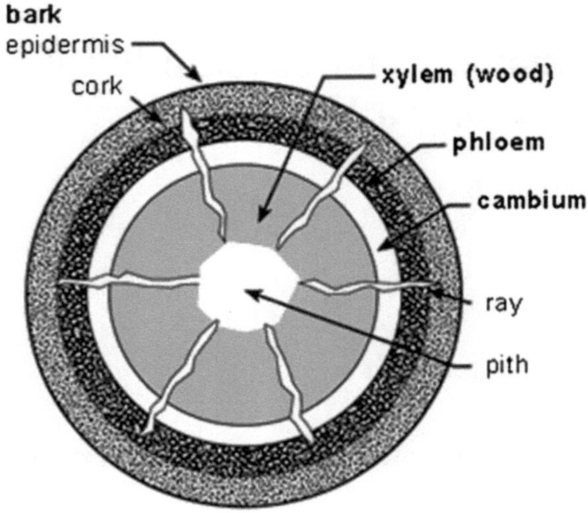

Fig. 2.1 Cross-section of a woody stem

Table 2.1 Basic chemical composition of wood

Element	% Dry weight
Carbon	49
Hydrogen	06
Oxygen	44
Nitrogen	Slight amount
Ash[a]	0.2–1.0

[a]What remains of wood after complete combustion in the presence of abundant oxygen
Source Zylkowski (2002)

as conifers or gymnosperms) are cone bearing trees which tend to have needle-like or scale-like evergreen leaves. Hardwood trees (deciduous trees or angiosperms), on the other hand, usually have broadleaves that are shed annually during the drier period. However, the technical terms ‘softwood’ and ‘hardwood’ do not have any bearing on the strength and/or toughness of a particular wood species. Some softwoods are harder than some hardwoods and vice versa.

The cellulose contents of both hardwoods and softwoods are about the same, but hardwoods generally have lesser amount of lignin than softwoods. Hardwoods also have a more complex cellular structure, more specialized tissues and are generally less uniform in structure than softwoods. For example, only the longitudinal cellular elements known as tracheids are responsible for both structural support and water conduction in softwoods. In hardwoods, tube like elements referred to as vessels facilitate the rise of water in the stem, while fibres provide structural support.

Table 2.2 Elemental composition of selected Nigerian hardwood species

Hardwood species	% Mass of element in oven-dry wood				
	Carbon	Hydrogen	Nitrogen	Oxygen	Ash
Acacia albida	46.53	6.34	0.48	41.23	3.33
Afrormosia laxiflora	45.65	7.37	0.91	43.58	2.49
Albizia adianthifolia	45.63	6.46	1.00	45.58	1.33
Albizia ferruginea	45.30	6.46	0.58	44.66	3.00
Albizia zygia	45.88	6.43	0.57	45.49	1.33
Annona senegalensis	46.01	7.11	0.72	44.00	2.16
Anonidium mannii	43.34	6.55	0.65	47.46	2.00
Anthocieista djalonensis	45.01	6.54	0.48	46.97	1.00
Berlinia grandiflora	44.80	6.63	0.67	43.57	4.89
Bombax costatum	45.70	6.19	0.51	45.6	2.00
Daniellia oliveri	45.73	6.51	0.81	44.62	2.33
Prosopis africana	47.84	7.20	1.63	37.68	5.65
Terminalia glaucenscens	45.54	6.31	0.52	45.63	2.00
Terminalia laxiflora	49.02	7.35	0.42	42.44	0.77
Pterocarpus erinaceus	47.71	6.30	0.54	44.45	1.00
Triplochiton scleroxylon	45.42	6.02	0.28	43.39	4.89

Source Lucas and Fuwape (1984)

Table 2.3 Organic compounds in wood (% oven-dry weight)

	Cellulose	Hemicellulose	Lignin
Hardwoods	40–44	15–35	18–25
Softwoods	40–44	20–32	25–35

Source Zylkowski (2002)

Hardwoods are found in both the tropics (i.e. the region of the earth near to the equator, between the Tropic of Cancer in the northern hemisphere and the Tropic of Capricorn in the southern hemisphere, also referred to as the tropical zone and the torrid zone) and temperate regions (i.e. tepid latitudes of the earth that lie between the tropics and the polar regions). The world's largest tropical rainforests are in South America, Africa and Southeast Asia, while the major temperate hardwood forests are in eastern North America, East Asia and Europe. Smaller temperate tropical forests also occur in Australasia and South America. Some of the popular temperate hardwoods include Oak, Maple, Beech, Poplar and Elm, while the major sources of tropical hardwoods in Africa include Cameroon, Cote D'ivoire, Ghana, Morocco and Nigeria. The physical and mechanical properties of tropical and temperate hardwood species are similar and in some cases comparable in some respects, but they are not all quite the same.

A tree is generally divided into areas of living tissues and those which are dead. The part of the wood of trees that contain living cells and water, store food and provide a degree of mechanical support is called the ***sapwood***. The core of dead tissue in the stem is called the ***heartwood*** (Fig. 2.2). While the heartwood does not function in the life processes of the standing tree, it provides a mechanical function of supporting the tree. The transformation of *sapwood* to *heartwood* is usually accompanied by a darkening of the heartwood tissue which tends to collect excess nutrients that metabolize into extractives such as waxes, oils, resins, fat and tannins along with aromatic and colouring materials. Hence, the sapwood is usually lighter in colour than the heartwood. However, it is not all woods that have a distinct colour difference between heartwood and sapwood.

Trees grow at different rates throughout the year, resulting in what is referred to as 'growth rings' in the wood. Under favourable growing condition, trees grow at a faster rate, resulting in lower density fibres. This portion of the wood is known as ***earlywood*** or ***springwood***. Wood formed in the less favourable growing season is referred to as ***latewood*** or ***summerwood*** (Fig. 2.3). In temperate regions, most hardwood species show sufficient difference in earlywood and latewood, as clearly observable in the annual growth rings, whereas no well-defined annual rings may be visible in hardwoods grown in the tropics where growth may be nearly continuous.

The three recognized principal planes in the stem of woody plants through which wood is customarily examined are transverse (X), also called the cross-section, radial (R) and tangential (T) planes or surfaces (Fig. 2.4). The transverse surface is exposed when wood is cut or sawn at right angles to the longitudinal axis of the tree stem, i.e. the surface presented at the end of a log. The radial and tangential surfaces of wood are at right angles to the transverse section. The radial surface is exposed when the cut follows a radius of a cross-section of the log, along the axis of the log. The tangential surface is exposed when the bark is peeled from a tree. The basic physical and mechanical properties of wood tend to vary along these three principal planes; hence, wood is said to be anisotropic.

All woods in general, regardless of its botanical origin, and hardwoods in particular possess the following characteristics:

Susceptibility to Bio-deterioration: As a biological material, wood contains a lot of moisture, especially when green (i.e. wet, either as a standing tree or freshly felled timber). Some species are very wet when cut, while others are comparatively dry. If the moisture in wood is not properly handled during timber harvesting and processing, there is the likelihood of incurring avoidable losses arising from:

- Bio-deterioration and the attendant material loss,
- Extra handling and transportation costs,
- Hampered processing efforts that may yield products of relatively poor quality and hence lower earnings.

Bio-deterioration of the wood material is a serious hazard when handling low-density timber species such as *Alstonia boonei*, *Antiaris africana* and *Ceiba pentandra* in the warm humid rainforests. Rapid removal of such logs from the forests after harvesting is therefore essential to minimize the chances of substantial

Fig. 2.2 Cross-section of a woody stem showing the pale sapwood and the dark heartwood

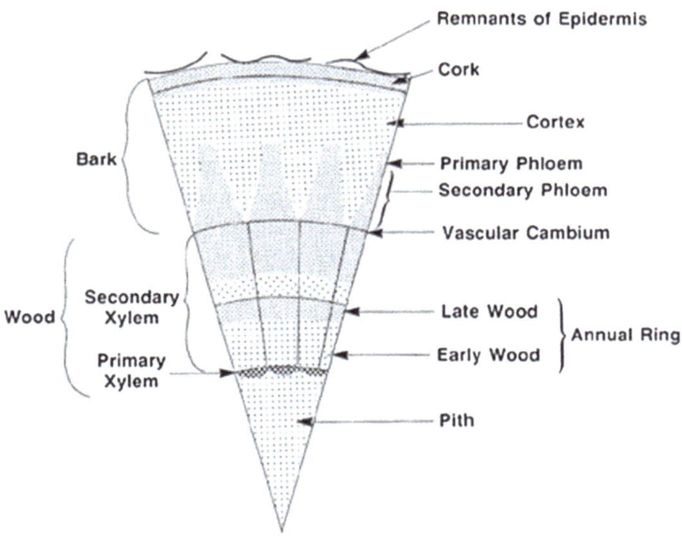

Fig. 2.3 Section of a woody stem showing early wood and late wood

biodegradation. However, contrary to the a widely expressed opinion, wood is generally a highly durable material, if treated properly, i.e. if it is not exposed to an environment characterized by high relative humidity (moisture), high oxygen

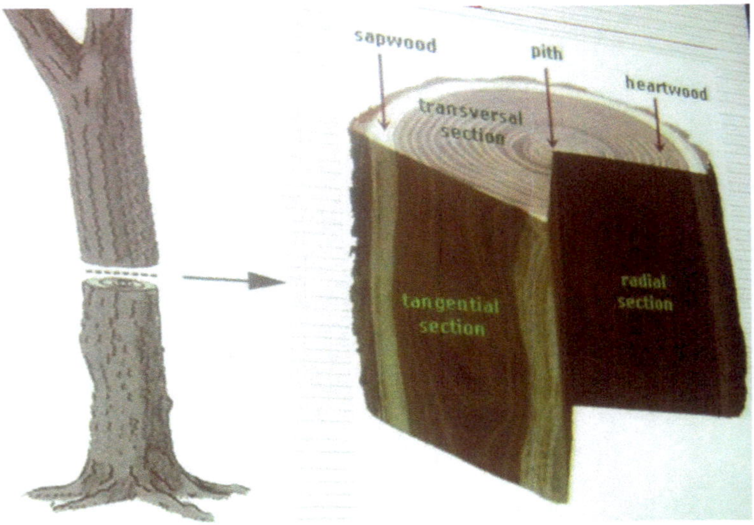

Fig. 2.4 Wood anatomical planes

supply and relatively high temperatures for prolonged periods. Nevertheless, some wood species (usually the denser ones including *Afzelia africana*, *Nesogordonia papaverifera* and *Milicia excelsa*) are more durable than others. Also, the heartwood is usually more durable than the sapwood. The good news is that many hardwood species can be treated with appropriate preservative chemicals to enhance their durability. Drying or seasoning also tends to improve their durability.

Variations in Physical Properties: Hardwoods are highly variable within and among species and from one portion of a tree to another with considerable variations in weight, diameter, density, strength and other properties, even in a metre run of wood. No two pieces of hardwood are exactly the same. There are also very wide variations in size and shape as shown in Table 2.4. Some hardwood saw logs are straight, others are fairly straight, while others are bent or crooked; some saw logs are big in diameter (or girth), while others are small. Irregularities in log sizes and shapes have significant effects on volume recovery (also known as conversion ratio or conversion efficiency) during sawmilling. On the basis of shape alone, the conversion ratio for relatively straight logs is usually higher, regardless of the species, than for fairly straight and crooked logs (the ratio typically obtained for fairly straight logs usually falls in-between). However, on the basis of shape and size, different results can be obtained. For instance, smaller logs could give higher conversion ratios if:

(i) The apparently very large logs contain high volumes of unsound material;
(ii) The very large logs contain more taper than the smaller logs; and
(iii) The opening face (i.e. the first cut) from a very large log is larger than that of a comparatively smaller log.

Table 2.4 Sizes and densities of selected tropical hardwoods grown in Nigerian

S/No	Botanical name	Trade name	Average diameter range (m)	Approximate density range at 12% m.c. (kg/m^3)
1.	*Mitragyna stipulosa*	Abura	0.6–0.9	510–640
2.	*Lovoa trichilioides*	Sida	0.9–1.2	460–640
3.	*Terminalia superba*	Afara	0.9–2.4	410–570
4.	*Lophira alata*	Ekki	1.2–1.5	1020–1201
5.	*Nauclea diderrichii*	Opepe	0.9–1.8	730–800
6.	*Ceiba pentandra*	Araba	1.5–1.8	210-360
7.	*Mansonia altissima*	Mansonia	0.6–1.0	580–720
8.	*Afzelia africana*	Apa	1–1.2	730–900
9.	*Khaya ivorensis*	Lagos mahogany	0.7–1.2	410–570
10.	*Pycnanthus angolensis*	Akomu	1–1.2	410–640
11.	*Gossweilerodendron balsamiferum*	Agba	1.5–2.5	460–570
12.	*Milicia excelsa*	Iroko	0.9–3	510–800
13.	*Entandrophragma cylindricum*	Sapele wood	1.5–1.8	580–720
14.	*Entandrophragma candollei*	Omu	1.5–2.5	650–720
15.	*Erythrophloeum guineense*	Erun	1–1.5	910–1010
16.	*Alstonia congensis*	Ahun	0.9 1.0	330–450
17.	*Saccoglottis gabunensis*	Atala	0.8 1.2	810–1010
18.	*Piptadenia africana*	Agboyin	2.5 3.5	730–1,010
19.	*Nesogordonia papaverefera*	Oro	0.75 1.0	730–800
20.	*Triplochiton scleroxylon*	Arere	1–1.5	370–450
21.	*Khaya senegalensis*	Dry-zone mahogany	0.6–0.9	730–900

Source Okigbo (1964), Lucas (1983)

Barring these factors, the typical relationship of increased percentage of lumber recovery with increased log diameter typically holds.

Variations in Chemical Properties: Different hardwood species have varying chemical properties. For example, some contain resins (e.g. *Isoberlina doka*, *Isoberlina tomentosa* and *Klainedoxa gabonensis*), silica (e.g. *Cola gigantea*, *Maesopsis eminii*, and *Morus mesozygia*) and other crystalline substances which produce dulling effects on wood machinery and create felling and sawing difficulties that tend to add to processing costs. Also, the thick bark of some logs must

Table 2.5 Estimates of bark volume in selected Nigerian hardwood timbers

Species	% Volume of bark in tree[a]
Milicia excelsa	13
Terminalia ivorensis	10
Bombax buonoponensis	8
Mimusops djaye	7
Mimusops heckelii	7
Daniellia ogea	7
Nauclea diderrichii	7
Khaya ivorensis	6.5
Khaya senegalensis	6.5
Sterculia rhinopetala	6.5
Piptadeniastrum africanum	3.3
Terminalia superba	3.3
Triplochiton scleroxylon	3.3
Cordia millennii	2.0
Nesogordonia papaverifera	2.0

[a]The measurements were from mature trees
Source Lucas (1975)

be removed before sawing to minimize processing costs. Estimates of bark volume in selected tropical hardwoods are presented in Table 2.5.

Presence of Natural Defects: Hardwoods are seldom free from natural defects. Some of these defects become visible only during processing, e.g. brittle heart and shakes. Natural defects in general tend to lower the strength properties and economic value of wood, while some defects can actually enhance its aesthetic appeal if properly handled during processing. The common natural defects of significant consequence in structural design are:

- **Knots**: A knot is that portion of a branch that is incorporated in the bole of a tree. The effect of a knot on the mechanical properties of wood is primarily due to the interruption of continuity and change in direction of wood fibres around the knot. If knots are tight and small, they may be of little consequence. If they are large, loose or missing, they seriously reduce strength and appearance value of the piece.
- **Checks**: These are cracks across the growth rings that occur during seasoning. They reduce strength and increase the likelihood of splitting due to nailing near the end of a piece.
- **Shakes**: These are cracks along the growth rings. They occur in standing tropical hardwood trees as a result of differential stresses of various origins, including wind, shrinkage of the heartwood and unbalanced growth stresses. Ruptures which follow the growth rings are called ring or wind shakes, while those that cross the annual rings are called heart checks. Regardless of the cause, shakes produce the same results as checks. The amount of loss in logs containing shakes depends on the location and size of the ruptures.

- **Bark pockets**: These are holes or cracks traceable to insects or birds in many cases. In other cases, the causes are unidentifiable.
- **Mineral streak/stain**: This is a form of discoloration observable in some hardwoods due to unusual concentration of minerals.

Susceptibility to manufacturing defects: Hardwoods in general are highly susceptible to manufacturing defects. Care should therefore be taken in the selection of the processing equipment and methods to minimize losses that are associated with poorly manufactured wood products.

Susceptibility to weathering agents: Hardwoods are highly susceptible to weathering agents, especially sunlight and rain, which tend to change their physical characteristics, especially colour, and adversely affect their strength properties. Reaction wood (i.e. tension wood and compression wood) which is recognizable through the presence of eccentric growth rings may develop in certain parts of leaning trees and at upper or lower sides of branches due to the effect of wind forces. While compression wood develops on the underside of leaning branches and leeward side of softwood stems, tension wood develops on the upper side of leaning hardwood stems. In green condition, the tensile strength of tension wood is typically low, but increases and may exceed that of normal wood when air-dried.

Hygroscopicity: Wood in general is hygroscopic; i.e., it tends to either gain or lose moisture in a bid to blend with the atmospheric conditions of the environment in which it is placed. The gain and/or loss of moisture results in swelling and shrinkage, respectively, particularly if the wood has not been properly dried and leads to the appearance of aesthetic and market value-reducing deformations such as *warpage*, *cupping*, *bowing* in many wood products.

Grain: Technically, the word *grain* refers to the orientation of wood-cell fibres. It is quite different from figure, which describes the distinctive pattern that frequently results from various grain orientations. Wood grain can, therefore, be defined as the longitudinal arrangement of wood fibres or the pattern resulting from this. This pattern of fibres is usually visible in a cut surface of wood. The two basic categories of grain are straight and cross grain. Straight grain runs parallel to the longitudinal axis of the piece. Cross grain deviates from the longitudinal axis in two ways, spiral grain or diagonal grain. The amount of deviation is called the slope of the grain. However, in general terms, wood species are described in terms of having *straight grain* which runs in a single direction along the cut wood as already defined, a product of a straight growing tree; *cross grain* in which some cells grow out from the major growth axis of the tree; *spiral grain* which develops as the trunk of the tree twists in development; and *interlocked grain* which spirals around the axis of the tree, but reverses its direction for periods of years resulting in alternating directions of the spiral grain. The grain types exhibited by selected Nigerian hardwoods are presented in Table 2.6.

In describing the application of a woodworking technique to a given piece of wood, the direction of the technique may be:

Table 2.6 Selected tropical hardwood timbers, their specific gravities and grain type

Botanical name	Specific gravity at 12% moisture content	Grain type
Afzelia africana	0.82	Interlocked (moderate)
Albizzia zygia	0.71	Interlocked (moderate)
Berlinia spp.	0.69	Interlocked (pronounced)
Brachystegia spp.	0.67	Interlocked (pronounced)
Combretodendron macrocarpum	0.87	Interlocked (moderate)
Cylicodiscus gabunessis	0.93	Interlocked (moderate)
Daniellia ogea	0.51	Straight
Distemonanthus benthamianus	0.67	Interlocked (pronounced)
Entandrophragma angolense	0.56	Interlocked (moderate)
Entandrophragma cylindricum	0.67	Interlocked (moderate)
Entandrophragma utile	0.64	Interlocked (pronounced)
Erythrophloeum guineense	0.83	Interlocked (moderate)
Gossweilerodendron balsamiferum	0.50	Interlocked (moderate)
Guarea spp.	0.63	Straight
Holoptelia grandis	0.64	Interlocked (moderate)
Isoberlinia doka	0.72	Interlocked (moderate)
Khaya spp.	Variable	Interlocked (moderate)
Lophira alata	1.19	Interlocked (pronounced)
Lovoa trichilioides	0.56	Interlocked (moderate)
Mansonia altissima	0.64	Straight
Milicia excelsa	0.66	Interlocked (moderate)
Mimusops djave	0.61	Interlocked (pronounced)
Mitragyna cilicta	0.54	Interlocked (moderate)

(continued)

Table 2.6 (continued)

Botanical name	Specific gravity at 12% moisture content	Grain type
Nauclea diderrichii	0.75	Interlocked (moderate)
Nesogordonia papaverifera	0.80	Interlocked (moderate)
Piptadeniastrum africanum	0.69	Interlocked (pronounced)
Pterygota macrocarpa	0.69	Interlocked (pronounced)
Scottellia coriacea	0.64	Straight
Sterculia oblonga	0.77	Interlocked (moderate)
Sterculia rhiopetala	0.80	Interlocked (moderate)
Terminalia ivorensis	0.54	Interlocked (moderate)
Terminalia superba	0.45	Straight
Triplochiton scleroxylon	0.37	Straight

Source Lucas (1985)

- with the grain (easy, giving a clean result)
- against the grain (heavy going, giving a poor result such as chipping or tear-out)
- across the grain (direction of cut is across the grain lines, but the plane of the cut is still aligned with them)
- end grain (at right angles to the grain, for example trimming the end of a plank).

Grain alignment must also be considered when joining pieces of wood or designing wooden structures. For example, a stressed span is less likely to fail if tension is applied along the grain, rather than across the grain. Grain direction will also affect the type of warping seen in the finished item. Interlocked grain occurs very commonly in tropical hardwoods and only occasionally in hardwoods of the temperate regions. However, all grain types except straight grain can be a blessing or a curse. Because wood with anything other than straight grain may be sawn to produce sometimes exquisite figure, errant grain becomes a blessing. In structural applications, such as home construction, lumber with anything other than straight grain loses some strength—since the presence of interlocked grains decreases the static bending strength and stiffness in bending of structural members—and hardwood boards without straight grain require extra care in machining to avoid tear-out and other reactions. On the other hand, interlocked grain tends to increase the resistance of lumber to splitting. Some woods are also chosen for a particular application because of the appearance of their grain. Naturally finished furniture, wall panelling and siding are typical examples.

2.2 Physical Properties of Hardwoods

Physical properties of wood are those properties that do not require mechanical stress for their determination. The fundamental physical properties of hardwoods that are of interest in structural design include the following.

2.2.1 *Density*

The density of a piece of wood is determined by the amount of wood substance present per unit volume. It is a measure of the amount of cell wall material relative to the size of the cell cavity. It is therefore related to both the relative proportions of the different types of cells present and the absolute wall thickness of any one type of cell. Wood density is significantly influenced by the presence of both moisture and extractives. The presence of moisture increases the mass of the wood and also tends to induce swelling, thereby affecting its volume. Therefore, the mass and volume of the wood must be determined at the same moisture content, and the density so obtained must be specified as being the derived value at that particular moisture content. Generally, mass and volume are determined at zero per cent moisture content, usually following a period of oven drying at 103 ± 2 °C. However, density is frequently quoted at 12% moisture content, since wood at this moisture content is in equilibrium with a relative humidity of 65%. Although the density of wood can vary extensively, the density of the actual cell wall material is constant at about 1500 kg/m^3; i.e., the maximum oven-dry density of wood is 1500 kg/m^3.

The densities of selected tropical hardwoods are presented in Table 2.4. Examples of light-to-medium density tropical hardwood timbers (density = 60–800 kg/m^3) include *Afzelia Africana, Alstonia boonei, Antiaris Africana, Brachystegia nigerica, Ceiba pentandra, Lovoa trichiloides, Nesogordonia papaverifera, Melicia exelsa, Ricinodendron heudelotii, Terminalia superba, Pycnanthus angolense* and *Triplochiton scleroxylon*. The heavy species (density > 800 kg/m^3) include *Anogeissus leiocarpus, Erythrophloeum ivorense, Nauclea diderrichii* and *Lophira alata*.

2.2.2 *Specific Gravity*

The specific gravity (relative density) of wood is the ratio of the density of wood to the density of water at 4 °C. This value also varies with the moisture content. For engineering applications, the specific gravity is frequently presented as the ratio of oven-dry mass to volume at 12% moisture content. The relationship between density and specific gravity of wood is represented by the equation:

$$\boldsymbol{\rho}_{\boldsymbol{\mu}} = \mathbf{G}_{\boldsymbol{\mu}}(1 + 0.01\boldsymbol{\mu})\boldsymbol{\rho}_{\mathrm{w}}$$

where

ρ_μ = Density at moisture content μ
G_μ = Specific gravity at moisture content μ
ρ_w = Density of water.

Specific gravity is an excellent index of the amount of wood substance present in a piece of wood and a good index of the mechanical properties of clear, straight-grained wood that is free from defects. However, specific gravity values also reflect the presence of gums, resins and extractives, which contribute little to mechanical properties. The relationship between specific gravity and mechanical properties of wood is represented by the equation:

$$\mathrm{P} = \mathrm{kG^n}$$

where

P = Any mechanical property
K and n = Constants that depend on the specific mechanical property and species
G = Specific gravity

The specific gravities of selected tropical hardwood timbers are shown in Table 2.6.

2.2.3 *Moisture Content*

Green wood, which refers to wood in the growing tree or in freshly felled timber, contains a very large quantity of sap which is mainly water. Moisture is present in green wood in two forms: 'free water' that fills or partially fills the cell cavities of the timber and 'bound water' that is absorbed in the cell walls. The amount of moisture absorbed in the cell walls at fibre saturation point (FSP) is limited and is roughly the same for all timbers, varying from one-quarter to one-third (i.e. 25–30%) of the oven-dry weight of the wood. The remaining moisture is free water. The moisture content of wood is computed in terms of the percentage of the weight of water present relative to the dry weight of the wood, i.e.:

$$M.C(\%)\frac{W - D}{D} \times 100$$

where W is the initial wet weight, and D is the dry weight.

Table 2.7 Typical green moisture contents of selected Nigerian hardwoods

Species	Green moisture content (%)
Ricinodendron heudelotii	160
Ceiba pentandra	140
Guarea cedrata	100
Terminalia superba	85
Piptadeniastrum africanum	85
Alstonia booneii	80
Celtis zenkeri	80
Khaya ivorensis	70
Triplochiton scleroxylon	65
Nauclea diderrichhii	65
Gossweilerodendron balsamiferum	55
Cylicodiscus gabunensis	55
Brachystegia nigerica	50
Lophira alata	50

Source Mackay (1946)

The sapwood of softwoods contains more water than the heartwood, and in some species, the ratio can be up to 5:1. The difference is much smaller in hardwoods. Also, moisture content tends to decrease with height in a standing tree, while some species tend to develop localized regions of extremely high moisture content throughout the stem. These regions, known as 'water pockets', often create difficulties in drying wood to uniform moisture content. The moisture content of tropical hardwoods in the green condition varies largely with the species and growth factors. However, it is fairly constant for any one species and does not vary greatly throughout the season. Some species are very wet when cut, and others are comparatively dry. The moisture content can vary from as high as 160% to as low as 50% of the oven-dry weight as shown in Table 2.7.

Moisture content affects various other wood properties such as durability, strength, electrical resistance, thermal conductivity, machining and finishing, among others.

2.2.4 Nail-Holding Ability

Since nailing still remains the most commonly used fastening method for wood members in many parts of tropical Africa, good nail-holding properties are essential in wood used for construction. The ability of wood to hold nails is closely related to its density. Also, nail-holding resistance for hard woods is typically greater than for softer woods but woods that are so hard that they tend to split when nailed are likely to lose much of their holding ability. Quite a number of tropical hardwoods, e.g.

Daniellia oliveri, Distemonanthus benthamianus, Terminalia superba and *Terminalia ivorensis,* have good nailing properties. However, a number of other species, e.g. *Afzelia africana, Berlinia confusa, Berlinia gradiflora, Cola gigantea, Diospyros mespiliformis, Lophira alata, Nesogordonia papaverifera, Pentaclethra macrophylla* and *Scortellia coriace,* are difficult to nail, prone to splitting and should be pre-bored before nailing. Pre-boring a hole 75% of the nail diameter, the use of smaller nails or the use of blunt-pointed nails are some of the methods used to reduce the incidence of splitting.

2.2.5 Ease of Working

Softer hardwood species that have uniform grain are easiest to work (i.e., saw, shape, nail, etc) and are generally more resistant to splitting but usually they cannot be given a high polish. Wood that is easy to work increases labour efficiency and helps ensure uniformity in the strength of nailed joints. Resistance to cutting and working properties of selected tropical hardwoods are presented in Table 2.8.

2.2.6 Paint-Holding Ability

Hardwood species that have uniform grain and exhibit little swelling and shrinking; e.g., *Canarium schweinfurthii* and *Mimusops djave* hold paint well. Edge grain ordinarily holds paint better than flat grain. Softwood species in particular should be free of knots and excessive pitch if they are to be painted. Regardless of species, paint-holding properties are affected by moisture in the wood as well as exposure to sun and rain.

2.2.7 Odour

A number of timber species have distinctive odour traceable to infiltration products in the heartwood or the action of fungi, bacteria or moulds. In the latter case, both the sapwood and heartwood may have odour, and because of the starch deposits, the odour may be stronger in the sapwood. Apart from aiding in identification, odour plays a major role in the selection and mechanical processing of wood. Some tropical hardwoods have pleasant scents, e.g. *Entandrophragma cylindricum* and *Guarea cedrata*. Many species are free of odour. The obnoxious/offensive odour of some species, however, precludes them from industrial processing and utilization. For example, *Fillacapsis discophora* and *Mansonia altissima* may not be selected for particular applications such as the construction of certain types of storage facilities because of the pronounced odour. Apples and oranges in storage, for example, may absorb odours and flavour from a wood that has a strong odour.

Table 2.8 Wood working properties of selected tropical hardwoods

Trade name	Botanical name	Wood	Working properties
		Resistance in cutting[a]	General working qualities[b]
		VH, H, M, L, VL, S	E, G, F, P, VP
Abura	*Mitragyna ciliata*	M	G
Afara	*Terminalia superba*	L	G
African walnut	*Lovoa trichiloides*	M	G
Afrormosia	*Afrormosia elata*	M	F
Agba	*Gossweilerodendron balsamiferum*	L	G
Albizia	*Albizia* spp.	L	G
Alstonia	*Alstonia boonei*	VL	E/G
Anogeissus	*Anogeissus leiocarpus*	VH	F
Antiaris	*Antiaris africana*	L	G
Apa	*Afzelia africana*	H	F
Ayan	*Distemonanthus benthamianus*	M/H	G
Berlinia	*Berlinia* spp.	M/H	F/G
Camwood	*Pterocarpus soyauxii*	M	G
Canarium	*Canarium schweinfurthii*	M/H	G
Ceiba/Bombax	*Ceiba pentandra and B. buonopozense*	VL	E
Celtis	*Celtis* spp.	M	F/G
Cordia	*Cordia millennii*	L/VL	E/G
Daniellia	*Daniellia oliveri*	M	F/G
Danta	*Nesogordonia papaverifa*	M/H	G
Diospyros	*Diospyros mespiliformis*	VH	F
Dahoma	*Piptadeniastrum africanum*	M/H	F
Doka	*Isoberlina doka*	H	F
Ebony	*Diospyros* spp.	VH	F
Ekki	*Lophira alata*	VH	F/P
Erimado	*Ricinodendron heudelottii*	VL	G
Erun/Missanda	*Erythrophloeum ivorense*	VH	F/P
Essia	*Combretodendron macrocarpum*	M	F
Gedu-Nohor	*Entandrophragma angolense*	M/H	G
Guarea, black	*Guarea thompsonii*	M	G
Holoptelea	*Holoptelea grandis*	M/H	G/F
Idigbo	*Terminalia ivorensis*	M/L	E/G
Ilomba	*Pycnanthus angolense*	L	F/G
Iroko	*Meilicia excelsa*	M	G
Lolagbola	*Oxystigma oxyphyllum*	M	G/F

(continued)

Table 2.8 (continued)

Trade name	Botanical name	Wood	Working properties
		Resistance in cutting[a]	General working qualities[b]
		VH, H, M, L, VL, S	E, G, F, P, VP
Mahogany, Benin	*Khaya grandifoliola*	M/H	G/F
Mahogany, dry zone	*Khaya senegalensis*	M	F/G
Mahogany, Lagos	*Khaya ivorensis*	M/H	F
Makore	*Mimusops djave*	H	F
Mansonia	*Mansonia altissima*	M	G
Obeche	*Triplochiton scleroxylon*	L	E/G
Odoko	*Scottellia coriacea*	M/H	F/G
Ogea	*Daniellia ogea*	M	F/G
Okan	*Cylocodiscus gabunensis*	VH	F/P
Okwen	*Brachystegia* spp.	M	F/G
Omu	*Entandrophragma candollei*	M	G
Opepe	*Nauclea diderrichii*	M/H	F/G
Pterocarpus	*Pterocarpus erinaceus*	M	F/G
Pterygota	*Pterygota macrocarpa*	L	F
Sapele	*Entandrophragma cylindricum*	M/H	F/G
Sterculia, brown	*Sterculia rhinopetala*	M/H	G
Sterculia, yellow	*Sterculia oblonga*	M/H	G
Teak	*Tectona grandis*		
Utile	*Entandrophragma utile*	M	G

Source Okigbo (1964), RMRDC (1994)
Legend
[a]*Resistance in cutting VH*—Very high, *H*—High, *M*—Medium, *L*—Low, *VL*—Very low
[b]*General Working Qualities E*—Excellent, *G*—Good, *F*—Fair, *P*—Poor, *VP*—Very poor

2.2.8 Colour

The natural colour assumed by a particular timber species is generally a function of the type of extractives present mainly in the heartwood. It varies from tree to tree, and there may be wide variations within one tree. In the absence of extractives, wood tends to assume a rather pale straw colour, characteristic of the sapwood. In some species, the heartwood and the sapwood may be colourless, whereas in other species colour is a major distinguishing factor between the heartwood and the sapwood. In some species, the colour is fairly evenly distributed throughout the heartwood, whereas others may exhibit considerable colour variations. Colour is a major factor in the selection of species for various engineering applications, particularly in plywood, furniture and flooring material production. However, wood colour may change with exposure to light.

Tropical hardwoods come in different colours as shown in Table 2.9 and Fig. 2.5a–f, and many of the species have attractive colours. The variety in colour ranges from nearly white (e.g. *Alstonia boonei, Scottellia coriacea* and *Celtis* spp.), through different shades of brown (e.g. *Khaya ivorensis* and *Lovoa trichiloides*), brownish red (e.g. *Khaya senegalensis*), greyish cream (e.g. *Pterygota macroparpa* and *Pterygota bequaertii*), pink (e.g. *Mitragyna spp.*, *Terminalia superba*, *Guarea cedrata* and *Gossweilerodendron balsamiferum*), darkish grey (e.g. *Mansonia altissima)*, to black (e.g. *Diodspyros* spp.). *Lovoa trichiloides* is one of the most attractive tropical hardwoods. Its golden brown colour is marked with black streaks, and the surface is distinctly lustrous. It has been identified as a good substitute for American Black Walnut and European Walnut.

2.2.9 Texture

Texture refers to the sizes, distribution and proportional volumes of the cellular elements of which wood is composed. The texture of wood is, therefore, a function of variations in size and uniformity in arrangement of its cells. The term is often used interchangeably with grain. Depending on the relative size and distribution of the cellular elements, all woods can be categorized as 'fine-', 'coarse-', 'even-' or 'uneven'-textured. Wood having small diameter cells is said to be fine-textured, while those having large diameter cells are said to be coarse-textured. Woods in which the distribution of cell types or sizes across the growth ring is uniform or those in which the thickness of the cell wall is fairly constant across the growth ring are said to be 'even-textured'. Those having variations in size and/or distribution of cell walls are said to be 'uneven-textured'.

Texture is a major factor defining the decorative appearance of wood. Tropical hardwoods that are coarse-textured include *Afzelia Africana*, Albizia zygia, *Albizia ferruginea*, *Lophira* alata, Terminalia *ivorensis* and *Milicia excelsa*. *Entandrophragma utile*, *Gossweilerodendron balsamiferum*, *Holoptelea* grandis and *Triplochiton scleroxylon* are even-textured, while *Entandrophragma cylindricum* and *Mimusops djave* are fairly fine-textured.

2.2.10 Figure

This is the most important property that affects the appearance of wood. It is a manifestation of four anatomical features: grain, growth rings, rays and knots.

Table 2.9 Colours of selected tropical hardwood timbers

Trade name	Botanical name	Colour
Abura	*Mitragyna ciliata*	Brown to pink
Afara	*Terminalia superba*	Yellowish to pink
African mahogany	*Khaya* spp.	Reddish
African walnut	*Lovoa trichiloides*	Yellowish to pink
Afrormosia	*Afrormosia elata*	Light brown
Agba	*Gossweilerodendron balsamiferum*	Light brown to pink
Alstonia	*Alstonia boonei*	Nearly white
Anogeissus	*Anogeissus leiocarpus*	Dull yellow to grey
Apa	*Afzelia africana*	Reddish
Ayan	*Distemonanthus benthamianus*	Yellow
Celtis	*Celtis* spp.	White to pale yellow
Dahoma	*Piptadeniastrum africanum*	Chocolate brown
Danta	*Nesogordonia papaverifa*	Reddish brown
Ebony	*Diospyros* spp.	Black
Ekki	*Lophira alata*	Dark brown
Erun/missanda	*Erythrophloeum ivorense*	Dull brown with reddish dark streaks
Essia	*Combretodendron macrocarpum*	Reddish brown with decorative dark streaks
Gedu-Nohor	*Entandrophragma angolense*	Reddish
Guarea	*Guarea* spp.	Reddish
Guarea	*Guarea cedrata*	Pink
Holoptelea	*Holoptelea grandis*	Light to pale yellow
Idigbo	*Terminalia ivorensis*	Yellow
Ilomba	*Pycnanthus angolense*	Pinkish brown
Iroko	*Milicia excelsa*	Golden-orange to light brown
Lolagbola	*Oxystigma oxyphyllum*	Pale yellow
Mahogany, dry zone	*Khaya senegalensis*	Brownish red
Mahogany, Lagos	*Khaya ivorensis*	Pinkish brown to deep red
Makore	*Mimusops djave*	Dark brown to Reddish brown
Mansonia	*Mansonia altissima*	Varies, purple when fresh, fading to light fawn on exposure
Obeche	*Triplochiton scleroxylon*	Creamy-white to pale yellow
Odoko	*Scottellia coriacea*	White to pale yellow

(continued)

Table 2.9 (continued)

Trade name	Botanical name	Colour
Ogea	*Daniellia ogea*	Light brown usually marked with dark brown streaks
Okan	*Cylocodiscus gabunensis*	Alternate bands of reddish brown and greenish brown
Okwen	*Brachystegia* spp.	Pale fawn to fairly dark brown
Omu	*Entandrophragma candollei*	Dark brown
Opepe	*Nauclea diderrichii*	Light yellow with reddish or greenish streaks to deep gold
Pterocarpus	*Pterocarpus erinaceus*	Cream to dark brown
Pterygota	*Pterygota macrocarpa*	Greyish cream
Sapele	*Entandrophragma cylindricum*	Reddish brown
Sterculia, brown	*Sterculia rhinopetala*	Deep reddish brown
Utile	*Entandrophragma utile*	Reddish brown

Source Okigbo (1964)
Note The colours given are approximations for newly cut heartwoods. They may vary from tree to tree, and there may be wide variations within one tree. Colours change with exposure to light, and all timbers become grey after exposure to weather

Colour and figure give wood its aesthetic value for furniture, panelling, flooring and many other structural applications. Tropical hardwoods that have excellent figure include *Khaya ivorensis* (used for building elegant sports boats, cabins for luxury vessels, panelling railway coaches and aircrafts in the past), *Entandrophragma cylindricum* (used for panelling, furniture, staircases and window frames) and *Entandrophragma utile* (used for interior furniture, joinery and other construction works).

2.2.11 Luster

This is the property of wood that enables it to reflect light. It is distinct from colour and the ability of wood to accept a finish. Since woods without luster can be satisfactorily finished to a beautiful condition, it is therefore a relatively unimportant characteristic. Lustrous tropical hardwood species include *Pterocarpus soyauxii* and *Baphia nitida*.

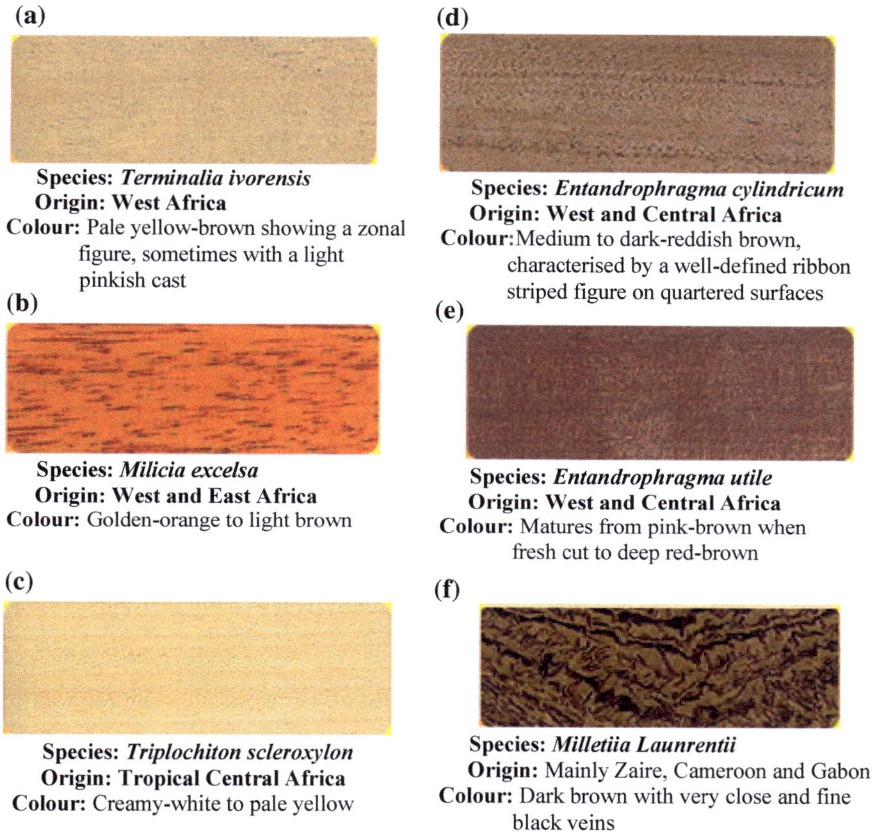

Fig. 2.5 a–f Tropical hardwoods of different colours

2.3 Concluding Remarks

There are different sources of hardwood supply in many timber-rich African countries. These include forest reserves, taungya farms and natural forests consisting of secondary forests, farmlands and derived savanna. Species such as *Milicia excelsa*, *Khaya ivorensis* and *Triplochiton sscleroxylon* are very common in high forests, *Khaya sengalensis* trees are very common in the savanna areas, while *Terminalia superba*, *Antiaris africana* and other lesser-used species are commonly found in the secondary forests. However, tropical hardwoods are presently not listed on the international stock exchange such as coffee and cocoa because of the lack of a generalized standard on the qualities of wood products and the fact that there are various species of wood that are suitable for the same end-use.

Also, consumer concerns about tropical deforestation have, in recent times, become a major factor that shapes the demand for tropical hardwood products in the international market. For example, negative consumer attitudes to the non-sustainable manner in which tropical forests are being exploited have weakened the demand for tropical timbers in many North western European countries in the last 20 years. There has been a variety of environmental and trade restrictions on production and exports in developed countries that impact on international trade patterns. The most common forms of environmental regulations that have international trade impacts include banning of imports or products, restricting selling and exporting of forest products, regulating method of production or processing (e.g. regulations banning or controlling certain timber preservation processes and materials) and regulation of product standards relating to the characteristics of the good.

Many developed countries have adopted quantitative restrictions to limit the import of 'unsustainably produced' forest products or to impose countervailing duties on imported products that benefit from an 'environmental' export subsidy, i.e. unsustainable forest management that leads to lower harvesting costs and thus lower export products. Some countries have also adopted the use of 'ecolabelling' and 'green certification[1]' to distinguish sustainably produced forest products or to ensure that forest product imports conform to domestic environmental standards and regulations. The main emphasis that was on timber and timber products in the early 1990s has, in recent years, been expanded to include pulp and paper.

Practice Questions

1. List and comment on five physical properties of hardwoods that are relevant for structural design.
2. As a consultant for a project involving the design and construction of wooden classroom furniture, list and describe five physical properties of wood you would take into consideration in recommending specific hardwood species for the project.
3. If you were to be appointed as a consultant for a project involving the design and construction of a timber-framed structure for use as temporary farm produce storage shed, describe five physical properties of hardwoods you would take into consideration in recommending specific species for your structural design.
4. Briefly describe the effects of moisture content and density on the structural utilization of tropical hardwoods.

[1]Timber certification is a process which results in a written statement (a certificate) attesting to the origin of wood raw material and its status and/or qualifications, often following validation by an independent third party. Its objective is to link the consumer who wishes to favour environmentally and/or socially responsible products with the producers of these products and the raw materials from which they are made.

References

Barbier EB (1995) Trade in timber-based forest products and the implications of the Uruguay round. Unasylva, Int J For For Ind 46(183):3–10

Bourke IJ (1995) International trade in forest products and the environment. Unasylva, Int J For For Ind 46(183):11–17

Burkill HM (2000) The useful plants of West Tropical Africa, 5. Royal botanic gardens, Kew, pp 296–297

Dinwoodie JM (1981) Timber, its nature and behaviour, international student edition. Van Nostrand Reinhold, New York, USA

Lucas EB (1975) Wasted tree products in Nigeria: Part I. Their causes extent and characteristics. Nigerian J For 5(1 & 2):24–30

Lucas EB (1983) Factors preventing wider commercialization of Nigerian Trees. For Prod J 33 (5):64–68

Lucas EB (1985) Theoretical evaluation of the effect of interlocked grain in wood on its compression strength. Nigerian J For 15(1&25):23–33

Lucas EB, Fuwape JA (1984) Burning and related characteristics of forty-two Nigerian Fuelwood species. Nigerian J For 14(1 & 2):45–52

Mackay JH (1946) On the utilization of forest products in Nigeria. Ph.D. thesis, University of Edinburgh

Okigbo L (1964) Some Nigerian woods, 2nd edn. Published by the Federal Ministry of Information, Lagos, Nigeria

Okigbo L (1965) A guide to building timbers in Nigeria, a monograph published by the Federal Department of Forest Research. Ibadan, Nigeria

Olorunnisola AO (2013) Harnessing the forester's harvest for sustainable development. An inaugural lecture delivered on May 16, 2013. Published by the publishing house, University of Ibadan, p 102

Panshin AJ, de Zeeuw C (1980) Textbook of wood technology. McGraw hill book company, New York

RMRDC (1994) Raw materials sourcing for manufacturing in Nigeria, 2nd edn. A report published by the Raw Materials Research and Development Council (RMRDC), Lagos, Nigeria

Walker JCF, Butterfield BG, Langrish TAG, Harris JM, Uprichard JM (1993) Primary wood processing principles and practice. Chapman and Hall Publishers, London, U.K

Zylkowski S (2002) Introduction to wood as an engineering material. In: Williamson TG (ed) APA Engineered Wood Handbook: 1.1–1.25. McGraw-Hill Publishers, New York, USA

References

Chapter 3
Mechanical Properties of Wood

Mechanical properties of wood are those that require mechanical stress for their determination. As already noted, the three principal planes recognised in the stem of woody plants in which wood is customarily examined are the longitudinal, radial, and tangential planes. Figure 3.1 shows the three principal axes of a piece of lumber/sawn wood with respect to grain direction and growth rings. The basic physical and mechanical properties of wood tend to vary along these three principal planes; hence wood is said to be anisotropic.

The two primary mechanical properties of wood are strength and toughness. Strength is the ability of wood to withstand stresses without appreciable deformation, while toughness is the ability of wood to withstand shock-loading. Even when loaded to the breaking point, members made from a tough species resist separation. Joists and beams under drive floors that will carry trucks and tractors need to be tough.

3.1 Strength of Wood

It is impossible to ascribe a single value to the strength of a piece of wood since it is an exceedingly variable property that is influenced by many factors, including inherent wood properties associated with local growing conditions, soil type, elevation silvicultural practices, and manufacturing techniques. Some of the strong tropical hardwood species include *Afzelia africana, Anogeissus leiocarpus, Distemonanthus benthamianus, Piptadeniastrum africanum, Nesogordonia papaverifera, Milicia excelsa, Erythrophloeum ivorense, Erythrophloeum guineense, Nauclea diderrichii, Pterocarpus erinaceus, Combretodendron macrocarpum* and of course, *Lopira alata*—one of the strongest timbers in the world. Other factors that influence the strength of wood and wood products include:

A.O. Olorunnisola, *Design of Structural Elements with Tropical Hardwoods*,
DOI 10.1007/978-3-319-65343-3_3

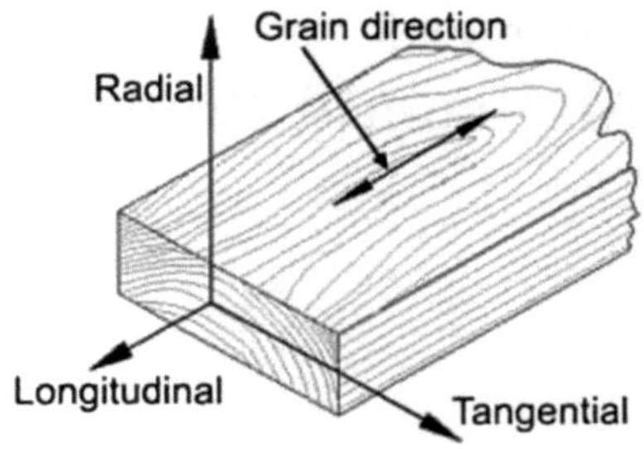

Fig. 3.1 Planes of reference in wood

i. **Rate of loading**:

The more rapid the rate of loading, the higher the apparent strength of wood. Static strength tests are typically conducted at a rate of loading or rate of deformation to attain maximum load in about 5 min. Higher values of strength are obtained for wood specimens loaded at a more rapid rate and lower values are obtained at slower rates. For example, the load required to produce failure in a wood member in 1 s is approximately 10% higher than that obtained in a standard static strength test. Over several orders of magnitude of rate of loading, strength is approximately an exponential function of rate of loading.

ii. **Duration of loading**:

Wood under stress will progressively lose strength with increasing time. This phenomenon is referred to as 'duration of load effect', 'static fatigue', or 'creep rupture'. The load that a wood product will support continuously over a long period of time is less than the load it will carry over a short period. For example, when continuously loaded for 10 years, a wood product will carry about 60% of the load that it will carry in the 5–10 min that it typically takes to test the strength of the same material in the laboratory. The allowable stress design method most commonly used for wood products is based on a 10 year duration of load. Published design stresses for wood also take this time effect into account. Table 3.1 presents the load duration modification factors applicable for allowable stress design.

iii. **Mode of applied stress**:

If wood, for example, is subjected to repeated loading over a relatively long time period, i.e., cyclic application of stress or fatigue, it will fail at a stress lower than its corresponding short-term ultimate strength. The rate at which the reduction in strength takes place is influenced by the level of maximum stress applied, the ratio of minimum to maximum stress applied, and the moisture content. The number of cycles to failure increases significantly with both decreasing moisture content and stress.

iv. **Direction of applied stress**:

The strength of wood is very sensitive to the orientation of the fibres relative to the direction of applied stress. The level of sensitivity varies with mode of stressing, being particularly high in the case of axial tension, i.e., wood is at its strongest

Table 3.1 Relationship between load duration and wood strength

Cumulative load duration	[a]Modification factor	Typical design loading situations	Design load
Permanent	0.9	Dead load	0.9 × Normal load
10 years	1.0	Occupancy live load	1.0 × Normal load
2 months	1.15	Snow load	1.15 × Normal load
7 days	1.25	Construction load or wind load in areas where hurricanes do not occur regularly	1.25 × Normal load
15 h	1.33	Wind load	1.33 × Normal load
10 min	1.6	Earthquake (seismic) load	1.6 × Normal load
2 s (Impact)	2.0	Impact load (very occasional)	2.0 × Normal load

[a]Also commonly referred to as adjustment factor
Source Wood handbook (Forest Products Laboratory 2010)

when stressed in axial tension and at its weakest when stressed in tension perpendicular to the grain. The relationship between grain angle and strength properties of wood has been empirically described by the Hankinson equation as follows:

$$N = \frac{PQ}{(P \sin^n \theta + Q \cos^n \theta)} \tag{3.1}$$

where

N = Strength at angle θ from fibre direction
P = Strength parallel to grain
Q = Strength perpendicular to grain
n = empirically determined constant, depending on strength property as shown in Table 3.2.

Table 3.2 Empirical constants for calculating mechanical properties of wood

Property	n	Q/P
Tensile strength	1.5–2	0.04–0.07
Compression strength	2–2.5	0.03–0.40
Bending strength	1.5–2	0.04–0.10
Modulus of elasticity	2	0.04–0.12
Toughness	1.5–2	0.06–0.10

Source Wood handbook (Forest Products Laboratory 2010)

Example 1 Use Microsoft Excel to compute the compressive strength properties of piece of lumber loaded at angles 0°, 15°, 30°, 45°, 60°, 75°, and 90°. Assume that the compressive strength parallel and perpendicular to the grain of the material are: 18 and 4.5 N/mm^2, respectively. Also, use Excel to plot a graph of the results.

Solution Steps

1. Open an Excel worksheet.
2. Input the compression parallel and perpendicular to grain values to be used in the computation into the Excel cell as shown in Fig. 3.2a.
3. Input the loading angles in degrees.
4. Use Excel to convert the angles into radians using the formula: (angle in degrees $\times$ π)/180. Input this formula using the appropriate Excel cell references as shown in Fig. 3.2a.
5. Display the Hankinson's formula.
6. Use Excel to perform the computation using the appropriate cell references as shown in Fig. 3.2a.

The result of the computation is shown in the excel worksheet presented in Fig. 3.2b.

To plot the results using Excel, you have to:

1. Highlight the loading angles and strength values as shown in Fig. 3.2c.
2. Click on 'INSERT' on the tools bar.
3. Click on 'Scatter diagram'
4. Click on 'Scatter with only markers'.

The result is shown in Fig. 3.2d.

Example 2 Write simple FORTRAN program to solve Hankinson formula

Solution

```
! a program to solve hankinson formula
real::t(100),tn(100),r(100),a,b,c,d,p,q
integer::f,i,kount,j
write(*,*)"dimensional array program for hankinson's formula"
kount = 0
write(*,*)"input the strength values of wood parallel and perpendicular to
grain, p, q"
read(*,*)p,q
write(*,*)"input number of loading angles being considered"
read(*,*)f
write(*,*)'input the angles of loading in degrees'
          do 100 i = 1,f
           read (*,*)t(i)
   tn(i) = t(i)*(3.142/180) !c to convert angle in radians to degree
   a = p*q
```

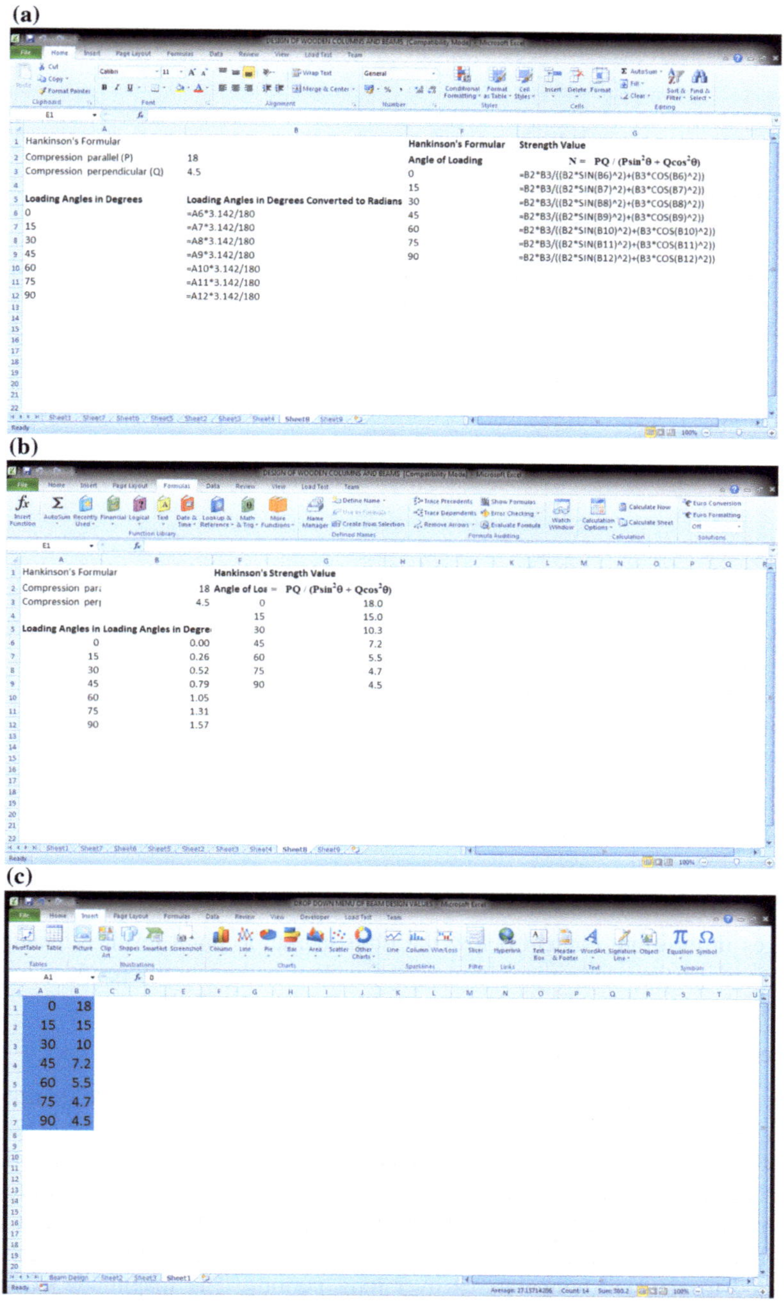

Fig. 3.2 **a** Excel worksheet showing the computations **b** Excel worksheet showing the results **c** Selection of values for plotting the graph using excel **d** An excel plot of strength values of wood at different loading angles

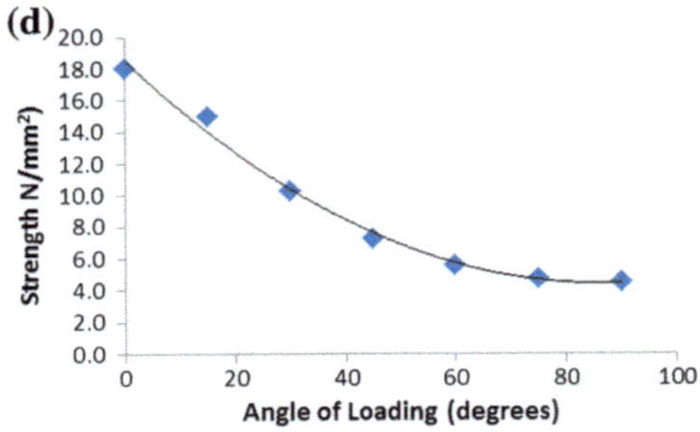

Fig. 3.2 (continued)

```
b = sin(tn(i))*sin(tn(i))
c = cos(tn(i))*cos(tn(i))
d = p*b
h= q*c
r(i) = a/(d+h)
   100 end do
   write(*,15)'s/no', 'angle', 'strength'
   15 format(a5,4x,a6,5x,a9)
   do 150 i = 1,f
     kount = kount +1
  write(*,20)kount,int(t(i)), r(i)
20 format(i2,8x,i2,10x,f6.2)
   150 end do
end
```

v. **Moisture Content**:

Changes in moisture content below the fibre saturation point result in changes in strength properties. A decrease in moisture content will generally result in increase in strength and vice versa as shown in Table 3.3.

vi. **Temperature**:

Temperature can generate both reversible (immediate) and irreversible (permanent) effects on the strength properties of wood and wood products in general. The strength of wood products tends to decrease as the temperature is increased, particularly if the material is relatively wet, since dry wood is less sensitive to temperature change than green wood. However, increases in temperature usually result in moisture content reduction. Permanent loss in strength would occur in wood subjected to elevated temperatures over long periods. The magnitude of loss in strength would depend on the temperature, duration of exposure, and wood moisture content among other factors.

Table 3.3 Approximate percentage increase/decrease in strength properties for 1% decrease/increase in moisture content

Property	% change per 1% change in moisture content
Static bending	
Fiber stress at proportional limit	4
Modulus of elasticity Modulus of rupture	2 4
Compression parallel to grain	
Fiber stress at proportional limit	5
Maximum crushing strength	6
Compression perpendicular to grain	
Fiber stress at proportional limit	5.5
Shear parallel to grain	
Maximum shearing strength	3
Tension perpendicular to grain	
Maximum tensile strength	1.5

Sources Panshin and de Zeeuw (1980), Zylkowski (2002)

vii. **Density**:

Increasing density results in a significant growth in all strength properties of wood, except axial tension.

viii. **Exposure to chemicals**:

The behaviour of wood and wood products when exposed to chemicals depends on a number of factors, the key ones being the pH and temperature of the chemical as well as the wood species involved. Chemicals tend to influence the strength properties of wood either by causing swelling (a reversible process), e.g., alcohol and other polar solvents; or by interacting with the wood at the chemical level (irreversible process).

3.2 Strength Properties of Wood Typically Considered in Structural Design

The strength properties of wood that are commonly measured and considered for design purposes include:

i. **Flexural/Bending Strength**:

This is the maximum stress sustained by a specimen loaded in a direction perpendicular to the grain. Flexural strength is also commonly referred to as modulus of rupture.

ii. **Tensile strength parallel to grain**:

This is the maximum tensile stress sustained in direction parallel to grain. Relatively few data are available on the tensile strength of various species of clear wood parallel to grain. In the absence of sufficient tension test data, the modulus of rupture is sometimes substituted for tensile strength.

iii. **Tensile strength perpendicular to grain**:

This is the resistance of wood to forces acting across the grain which tend to split a member. Values presented are the average of radial and tangential observations.

iv. **Compressive strength parallel to grain**:

This is the maximum stress sustained by a compression parallel-to-grain specimen having a ratio of length to least dimension of less than 11.

v. **Compressive strength perpendicular to grain**:

This is strength at proportional limit. There is no clearly defined ultimate stress for this property.

vi. **Shear strength parallel to grain**:

This is the ability of a wood member to resist internal slipping of one part upon another along the grain. Values presented are average strength in radial and tangential shear planes.

vii. **Rolling shear strength**:

This is the shear strength of wood when the shearing force is in a longitudinal plane and is acting perpendicular to the grain. Rolling shear strength is about the same in the longitudinal-radial and longitudinal-tangential planes.

viii. **Torsional strength**:

This is the resistance of a wood member to twisting about the longitudinal axis. For solid wood members, torsional shear strength may be taken as shear strength parallel to grain. Two-thirds of the value for torsional shear strength may be used as an estimate of the torsional shear stress at the proportional limit.

ix. **Impact bending**:

In the impact bending test, a hammer of given weight is dropped upon a beam from successively increased heights until rupture occurs or the beam deflects 152 mm (6-in.) or more. The height of the maximum drop, or the drop that causes failure, is a comparative value that represents the ability of wood to absorb shocks that cause stresses beyond the proportional limit.

The methods for determining these strength properties are available in different timber design codes and manuals.

3.3 Deformation of Wood Under Stress

The relationship between stress and the strain produced in all materials can exhibit any one of four characteristic behaviours illustrated in Figs. 3.3–3.6.

In each of the figures, σ = stress, ε = strain, and the slope of the curve relating stress to strain (symbolized by E, the modulus of elasticity) is the vital factor, where $E = \frac{d\sigma}{d\varepsilon}$.

In Fig. 3.3, $E = \alpha$ (infinity), i.e., there is no strain and the material is said to be rigid.

In Fig. 3.4; E = constant, and the stress–strain relationship is linear. The material is said to be elastic.

In Fig. 3.5, $E = f(\sigma)$. The material is said to be viscoelastic.

In Fig. 3.6, $E = 0$; there is no definite strain for any value of the stress and the material is said to be plastic.

Fig. 3.3 Rigid

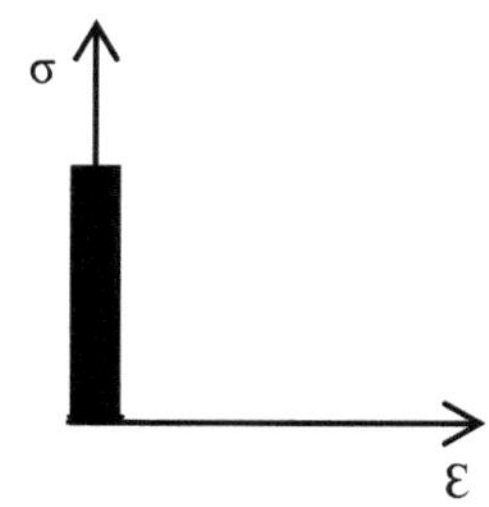

Fig. 3.4 Elastic

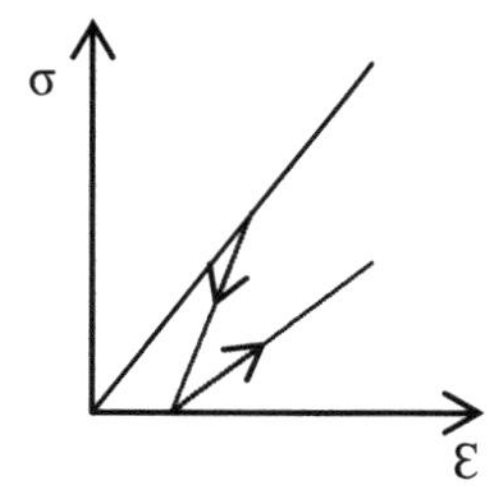

Fig. 3.5 Viscoelastic

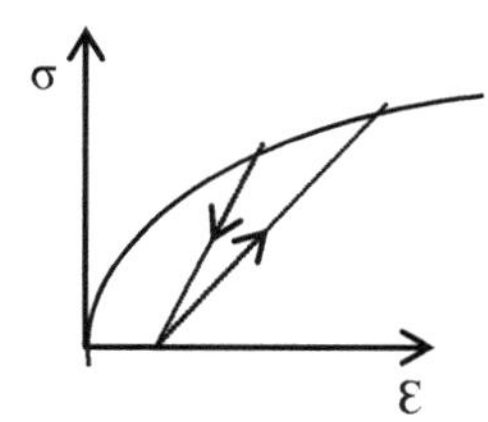

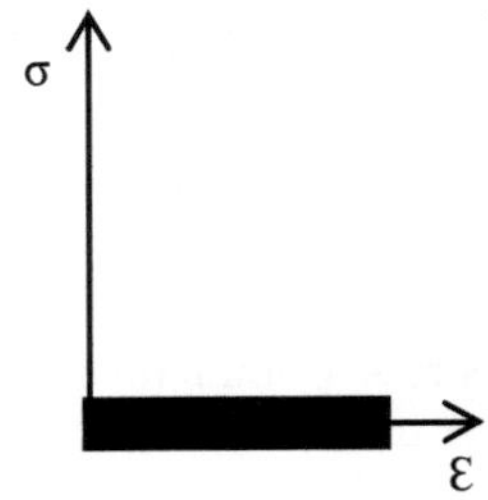

Fig. 3.6 Plastic

Materials which fracture when the strains are small are known as brittle, whilst materials which have an appreciable deformation before failure are said to be ductile. Real materials, including wood, generally have complex stress-strain relationships, but for purposes of analysis behaviour is simplified into:

(1) Behaviour as elastic material as shown by the single curve of Fig. 3.7.
(2) Behaviour as rigid material up to some yield stress, σy, after which it is plastic (Fig. 3.8).
(3) A combination of linear elastic and plastic behaviour (Fig. 3.9).

On loading, wood displays three forms of deformation behaviour—elastic, delayed elastic and viscous. Therefore, like so many of the materials, wood cannot be treated either as a truly elastic material, where, by Hook's law, stress is proportional to strain but independent of the rate of strain, or as a truly viscous liquid where, according to Newton's law, stress is proportional to rate of strain but independent of strain itself. Rather, wood is a viscoelastic material. Elastic deformation is, however, a better approximation of what happens to wood under longitudinal tensile loading than longitudinal compression. The approximation appears to approach

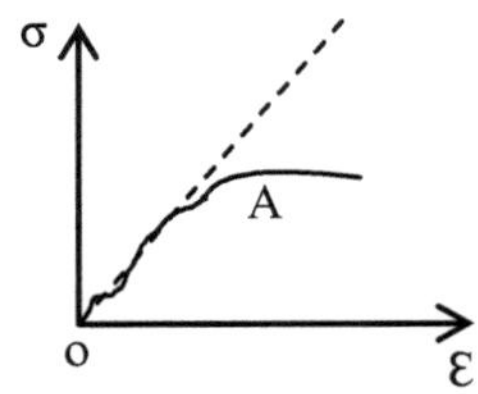

Fig. 3.7 Stress-strain curve for an elastic material

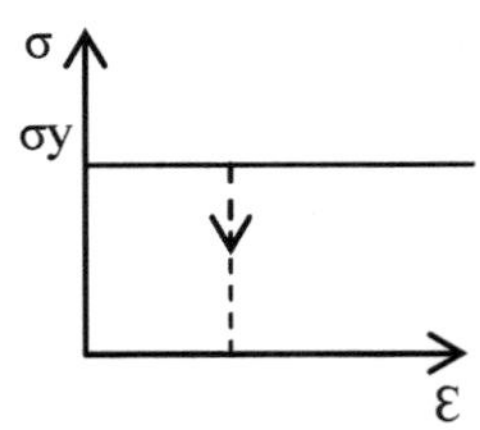

Fig. 3.8 Stress-strain curve for a rigid material

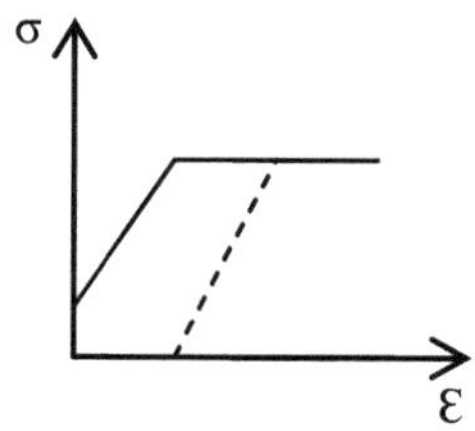

Fig. 3.9 Stress-strain curve for an elasto-plastic material

reality at the lower levels of longitudinal tensile or compressive loading up to the limit of proportionality, which generally occurs at about 60% of the ultimate load to failure in longitudinal tension, and between 30 and 50% of the failure value in longitudinal compression.

3.4 Modulus of Elasticity

The terms modulus of elasticity and stiffness, though not really having the same meaning, are sometimes used interchangeably. The stiffness of wood is its ability to resist deflection or bending when loaded. It is an important characteristic for wooden studs, joists and beams. Stiffness is necessary to prevent a gradual deflection in members which are loaded continuously over long periods of time. In general, the strongest species are also the stiffest. However, there are exceptions which are noted for their stiffness although they are only of medium strength.

Elasticity implies that deformations produced by low stress are completely recoverable after loads are removed. When loaded to higher stress levels, plastic deformation or failure occurs.

The modulus of elasticity in the longitudinal direction is one of the principal elastic constants of wood. Both static methods (a destructive method based on the application of direct stress and the measurement of resultant strains) and dynamic methods, (a non-destructive technique based on resonant vibration from flexural, torsional or ultrasonic pulse excitation) can be used to determine the elastic modulus. However, its determination from stress-strain curves remains a common method employed. Although the test is frequently conducted in the bending mode, it can also be performed in the compression, tension or shear modes. The values of the modulus in tensile, compressive and bending modes are approximately equal. In the three-point bending test (Fig. 3.10), measurement of deflection provides the static modulus using the equation:

$$E = \frac{PL^3}{48I\delta} \tag{3.2}$$

Fig. 3.10 A Wood specimen undergoing three-point bending test

where

E = Modulus of elasticity (N/mm^2)
P = Load applied to the centre of the span (N)
L = Span, i.e., distance between supports (mm)
I = Moment of inertia of the cross section (mm^4)
δ = Deflection at the centre of the span (mm).

The equivalent formula for four-point loading configuration shown in Fig. 3.11 is available in standard textbooks on wood structures as well as timber design codes. Using the appropriate rate of loading and testing equipment with acceptable degree of sensitivity, a curvilinear shaped load-deflection line would be obtained which could treated as linear by introducing a straight—line approximation. The approximation can take the form of either a tangent or secant as shown in Fig. 3.12. However, tangent lines are typically used for timber and wood fibre composites.

The three moduli of elasticity, which are denoted by E_L, E_R, and E_T, respectively, are the elastic moduli along the longitudinal, radial, and tangential axes of wood. These moduli are usually obtained from compression tests. The moduli of elasticity tend to vary within and between species and with moisture content and specific gravity.

Fig. 3.11 A Wood Specimen undergoing four-point bending test

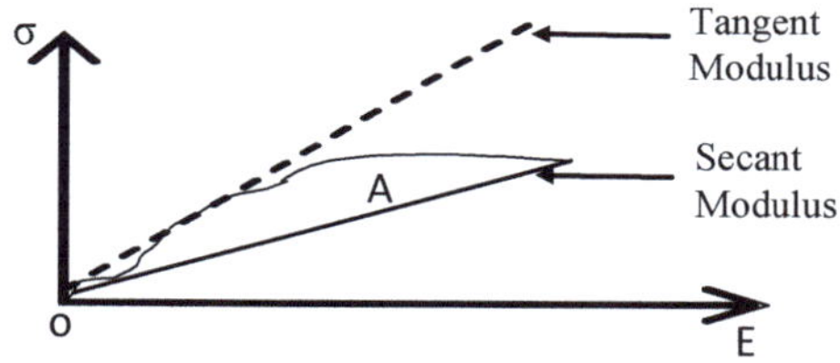

Fig. 3.12 Tangent and secant moduli approximation of deformation curve for wood

3.5 Modulus of Rupture

Modulus Of Rupture (MOR) is a parameter for measuring the bending strength of wood. It is the magnitude of the load required to cause failure in bending and it is influenced by wood density. Although MOR is an accepted criterion of strength, it is not a true stress because the formulas by which it is computed is valid only to the elastic limit, i.e.,

$$\text{MOR} = \frac{3PL}{2bd^2} \tag{3.3}$$

for three point bending test (rectangular cross section)

$$\text{MOR} = \frac{3PL}{4bd^2} \tag{3.4}$$

for four point bending test where the loading span is 1/2 of the support span (rectangular cross section)

$$\mathrm{MOR} = \frac{PL}{bd^2} \tag{3.5}$$

for four point bending test where the loading span is 1/3 of the support span (rectangular cross section)

where

P = Load at a given point on the load deflection curve (N)
L = Span, i.e., distance between supports (mm)
w = Width of the member (mm)
d = Depth of the member (mm).

3.6 Modulus of Rigidity

The modulus of rigidity, also referred to as shear modulus, indicates the resistance to deflection of a wood member caused by shear stresses. This property reflects the maximum load-carrying capacity of a wood member in bending and is proportional to maximum moment borne by the specimen. The three moduli of rigidity denoted by G_{LR}, G_{LT} and G_{RT} are the elastic constants in the *LR*, *LT*, and *RT* planes, respectively. For example, *GLR* is the modulus of rigidity based on shear strain in the *LR* plane and shear stresses in the *LT* and *RT* planes. As with moduli of elasticity, the moduli of rigidity vary within and between species and with moisture content and specific gravity.

3.7 Poisson's Ratio

When a wood member is loaded axially, the deformation perpendicular to the direction of the load is proportional to the deformation parallel to the direction of the load. The ratio of the transverse to axial strain is called Poisson's ratio. The Poisson's ratios are denoted by μ_{LR}, μ_{RL}, μ_{LT}, μ_{TL}, μ_{RT}, and μ_{TR}. The first letter of the subscript refers to direction of applied stress and the second letter to direction of lateral deformation. For example, μ_{LR} is the Poisson's ratio for deformation along the radial axis caused by stress along the longitudinal axis. Values for μ_{RL} and μ_{TL} are less precisely determined than are those for the other Poisson's ratios. Poisson's ratios vary within and between species and are affected by moisture content and specific gravity. Poisson's ratios for tropical hardwoods are not readily available in literature, while those for temperate hardwoods have been published in the popular Wood Handbook.

3.8 Creep and Relaxation

When initially loaded, a wood member deforms elastically. If the load is maintained, additional time-dependent deformation occurs. This is called creep. Creep occurs at even very low stresses, and it will continue over a period of years. For sufficiently high stresses, failure eventually occurs. At typical design levels and use environments, after several years the additional deformation caused by creep may approximately equal the initial, instantaneous elastic deformation. Ordinary climatic variations in temperature and humidity will cause creep to increase. An increase of about 28 °C in temperature can cause a two—to three-fold increase in creep. Green wood may creep four to six times the initial deformation as it dries under load.

Unloading a wood member results in immediate and complete recovery of the original elastic deformation and after time, a recovery of approximately one-half the creep at deformation as well. Fluctuations in temperature and humidity increase the magnitude of the recovered deformation. Relative creep at low stress levels is similar in bending, tension, or compression parallel to grain, although it may be somewhat less in tension than in bending or compression under varying moisture conditions. Relative creep across the grain is qualitatively similar to, but likely to be greater than, creep parallel to the grain. The creep behavior of all species is approximately the same.

If instead of controlling load or stress, a constant deformation is imposed and maintained on a wood member, the initial stress relaxes at a decreasing rate to about 60–70% of its original value within a few months. This reduction of stress with time is commonly called relaxation. As with creep, relaxation is markedly affected by fluctuations in temperature and humidity.

3.9 Hardness

This is the ability of wood to resist denting, scratching, and wear. It is generally defined as resistance to indentation using a modified Janka hardness test, measured by the load required to embed an 11.28 mm (0.444-in.) ball to one-half its diameter. Values presented are the average of radial and tangential penetrations. Although hard species are difficult to work and are subject to splitting, they are desirable for flooring, stair treads and bearing blocks used at the top of posts, which must be hard in order to avoid crushing. For products such as cabinets and furniture, hard wood is desirable because it will resist scratches and produce a high polish. *Lophira alata* is one of the hardest, strongest and most durable tropical hardwoods. It cannot be nailed without pre-boring.

3.10 Fatigue

Fatigue is defined as the progressive damage that occurs in a wood material subjected to cyclic loading. The loading may be repeated (stresses of the same type, e.g., compression or tension) or reversed (alternating compression and tension stresses). When sufficiently high and repetitious, cyclic loading stresses can result in fatigue failure. The number of cycles that are sustained before failure is a referred to as *fatigue life*, while the maximum stress attained in the stress cycle is referred to as *fatigue strength*. *Fatigue strength* is approximately exponentially related to *fatigue life*, i.e., fatigue strength tends to decrease approximately linearly as the logarithm of number of cycles increases. The *fatigue strength* and life of a piece of wood also depend on the frequency of cycling, repetition or reversal of loading, range factor (i.e., the ratio of minimum to maximum stress per cycle), temperature, moisture content, and specimen size. Negative range factors imply repeated reversing loads.

Fatigue strength is lower for wood containing small knots or a 1-in-12 slope of grain than for clear straight-grained wood and even lower for wood containing a combination of small knots and a 1-in-12 slope of grain; it is slightly lower in shear than in tension parallel to the grain; and relatively high in compression parallel to the grain compared with other properties.

3.11 Toughness

This is the ability of wood to withstand shock-loading. Even when loaded to the breaking point, members made from a tough species resist separation. Tough woods tend to deflect considerably before breaking. Even after fracturing, the fibres tend to hang together to resist separation.

Practice Questions

1. Comment on the effects of the following on the strength properties of wood:
   - (i) Rate of loading
   - (ii) Duration of loading
   - (iii) Grain angle
   - (iv) Moisture content
2. A *Milicea excelsa* piece of lumber is to be nailed at angle 30° to another piece of wood in a timber—framed construction. If the piece of lumber is to be subjected to compressive stress, determine: (a) Its compressive load bearing capacity in wet and dry conditions, using the Hankinson's formula (assuming the compression strength parallel to grain values are 7.1 N/mm^2 and 9 N/mm^2 in wet and dry conditions respectively; and the compression perpendicular to grain values are 2.5 N/mm^2 and 3.15 N/mm^2 in wet and dry conditions respectively.

(b) Repeat the computations assuming that the angle of loading is 60°.

3. A piece of lumber is nailed at an angle of 45^0 to another piece of lumber in a timber-framed construction. If the compression strength parallel to the grain of the nailed timber species is 10 N/mm^2, while the compression strength perpendicular to the grain is 1.2 N/mm^2, determine the compressive load bearing capacity of the joint.
4. Use Excel to compute the static moduli of elasticity and rupture of a 750 mm long 50 mm (*width*) × 50 mm (*depth*) lumber subjected to three-point bending load, if the maximum deflection is 5 mm at a load of 1.5 kN.

References

Dinwoodie JM (1981) Timber, its nature and behaviour, international student edition. Van Nostrand Reinhold, New York

Forest Products Laboratory (2010) Wood handbook—Wood as an engineering material, general technical report FPL-GTR-190. U.S. Department of Agriculture, Forest Service, Forest Products Laboratory. Madison

Hoyle RJ Jr (1972) Wood technology in the design of structures. Mountain press publishing company, Missoula

Kukreja CB, Kishore K, Chawla R (2007) Material testing laboratory manual for quality control. Standard publishers distributors, New Delhi

Lucas EB (1983) Factors preventing wider commercialization of Nigerian trees. For Prod J 33 (5):64–68

Marshall WT, Nelson HM (1978) Structures. Pitman Publishing Ltd, London

NCP (2005) Nigerian code of practice for structural timber design. Standards Organisation of Nigeria

Okigbo L (1964) Some Nigerian woods, 2nd edn. Federal Ministry of Information, Lagos

Olorunnisola AO (2013) Harnessing the forester's harvest for sustainable development. An inaugural lecture delivered on May 16, 2013. The Publishing House, University of Ibadan, p 102

Panshin AJC de Zeeuw (1980) Textbook of wood technology. McGraw Hill book company, New York

Walker JCF, Butterfield BG, Langrish TAG, Harris JM, Uprichard JM (1993) Primary wood processing principles and practice. Chapman and Hall Publishers, London

Zylkowski S (2002) Introduction to wood as an engineering material. In: Williamson TG (ed) APA engineered wood handbook. McGraw-Hill Publishers, New York. pp 1.1–1.25

Chapter 4
Hardwood Timber Seasoning and Preservation

4.1 Wood Moisture Content and Seasoning

Another term commonly used in place of timber seasoning is drying. Seasoning is perhaps a better word than drying, but both are synonymous. Seasoning is essentially a partial drying of the wood, and its progress can best be gauged by measuring the amount of moisture remaining in the wood at any time, i.e. the moisture content. Once felled, timber starts to dry. If it is not in contact with a moist body and is protected from rain, it dries very slowly if the weather is inclement, more quickly if the sun is shining and quite rapidly if there are strong winds. It eventually assumes a moisture content that is in equilibrium with the surrounding air, which may be as high as 20% in a humid environment as obtained during the rainy season in many parts of the tropics and as low as 6% in a hot dry climate as obtained in the northern parts of many African countries.

Drying commences with the evaporation of water from the wood. The free water in the cell cavities of the wood material is evaporated first, leaving only the absorbed bound water in the cell walls. Drying up to this point does not affect the properties of the timber, apart from reducing its weight, but further drying causes the removal of water from the cell wall which causes the wood to shrink. Timber should be protected from rain and from the ground and stacked so that air can circulate freely around all surfaces. Thus, the risks of twisting and cupping and attack by fungi and insects are minimized. In favourable conditions, thin softwoods can be air-seasoned in weeks, but in unfavourable conditions, some hardwoods require a year or more.

A.O. Olorunnisola, *Design of Structural Elements with Tropical Hardwoods*,
DOI 10.1007/978-3-319-65343-3_4

4.2 Reasons for Human Intervention Timber Drying

Since wood drying is inevitable, the main reason for human intervention in the process is to exercise some form of control so as to minimize all forms of degradation that usually accompanies drying. The various practical reasons for seasoning wood may be summarized as follows:

- To ensure that all shrinkage has occurred before the timber is used.
- To make the timber stronger since the strength of wood increases considerably as the wood dries below 30% moisture content except in its resistance to impact.
- To prevent warping and splitting and to ensure stability when put to use.
- To obtain a better surface finish, before gluing, painting or polishing since timber must be dry if it is to take paints and polishes.
- To make the material less susceptible to decay since timber that is below 20% moisture content will not decay, and prevent wood-staining fungi and pinhole borer infestation.
- To facilitate impregnation with certain preservatives, i.e. to allow penetration by preservatives if and/or when required. At 25% moisture content, good penetration can take place.
- To reduce the cost of transportation.

4.3 Basic Principles of Drying

The variations in timber drying rates and duration in different climates and drying conditions occur simply because the timber drying process is controlled by factors such as:

- Relative humidity, i.e. the drying rate is faster at lower relative humidity;
- Air temperature, i.e. the drying rate is faster at higher temperature;
- Airflow across the timber surface, i.e. the drying rate is faster when airflows across the timber surface;
- Wood density (more resistance to diffusion is experienced in denser wood than in lighter ones. Hence, the rate of drying is less with denser timbers);
- Thickness of the material being dried (the thicker the timber, the shallower the drying gradient, at least during the later stages of drying. Hence, the rate of drying is less with thicker timbers);
- Permeability of the wood (with impermeable timbers, the absorbed water is extremely hard to remove as the pits are completely blocked. Permeability is also a function of density. Permeable and impermeable timbers of similar densities should dry from fibre saturation point at about the same rate).

Drying of timber itself occurs as a two-stage process: the movement of moisture from the interior to the surface of the board and the evaporation of moisture from the surfaces to the moving air stream. These events take place concurrently, but it is essential that the rate of evaporation be controlled and in balance with the rate at which moisture moves to the surface. A convenient empirical approach is to assume that the time to dry timber is proportional to an exponential power of both density and thickness, respectively. The various drying elements are usually manipulated in a mechanical drying system to maintain this balance.

The key drying parameters that are of interest in the design of wood structures include the following:

(i) Fibre Saturation Point **(FSP)**

Moisture is present in green wood in two forms: 'free water' that fills or partially fills the cell cavities of the timber and 'bound water' that is absorbed in the cell walls. The point at which the cell wall of the wood material is completely saturated with water, and no moisture (free water) remains in the cell cavities, is defined as the fibre saturation point (FSP). The moisture content of the individual cell walls at FSP varies for different species of wood and generally falls between 25 and 30% of the oven-dry weight of the wood. However, for many purposes, the FSP is assumed to be 30% for all woods. When a piece of wood has dried to 15% moisture content, about half of the total possible shrinkage has taken place; when dried to 8% moisture content, nearly three quarters of the maximum possible amount has taken place.

The FSP is important to the structural engineer for the following reasons:

- All the physical and mechanical properties of wood are greatly affected by fluctuations in the quantity of water present in wood. In particular, large changes in these properties begin to take place at the FSP. The strength, stiffness and dimensional stability of wood are all related to its moisture content. Hence, if wood is dried before use, not only can higher strength values be used in design, but a more durable structure will result.
- A wood cell will not shrink until it reaches the FSP, and shrinkage causes problems associated with dimensional changes in installed wooden components such as floor parquet, door posts. The problems are frequently compounded by anisotropy in shrinkage, i.e., the degree of shrinkage is typically different on the three principal axes of wood. The greatest shrinkage occurs in the direction of the growth rings, about half of this amount in the radial direction, and very small amount (0.2% or less) in the longitudinal direction or along the grain. Each 3% of shrinkage either radially or tangentially is roughly equivalent to a decrease in width or thickness of 0.03 mm per mm. For example, if the tangential shrinkage of *Melicia excelsa* at 6% moisture content is 6.3%, the shrinkage per mm of width in a 200-mm flat-sawn board would be

$$6.3/3 \times 0.03 = 0.06\,\text{mm (approximately) for 1 mm}$$

For a 200 − mm wide flat − sawn board, this value would become

$$200 \times 0.06 = 12\,\text{mm}$$

Knowing the total shrinkage of a species at 0% moisture content, the percentage of shrinkage at any moisture content below 30% can be calculated by assuming that each 1% change in moisture content is equal to 1/30 of the total shrinkage. For example, assuming the total tangential shrinkage of a species dried to 0% moisture content is 7.9%. At 25% moisture, its tangential shrinkage would be 5/30 × 7.9, which is about 1.3%; at 12%, its tangential shrinkage would be 18/30 × 7.9 which is about 4.7%. These computations can be easily done with Excel as demonstrated in the worked example below.

Example 1 Assuming the total tangential shrinkage of a species dried to 0% moisture content is 7.9%. Use Microsoft Excel to compute its tangential shrinkage if the moisture content of the environment is 25%.

Solution Steps

1. Open an Excel worksheet.
2. Input the shrinkage value at 0% moisture content and the moisture content at which the new shrinkage value is to be computed into the Excel cell as shown in Fig. 4.1.
3. Input the formula for computing moisture content reduction (i.e. MCR = 30 − MCA, where MCR = Moisture content reduction, MCA = Moisture content at which shrinkage is to be computed).
4. Input the multiplier equation (i.e. Multiplier = MCR/30) using the appropriate Excel cell references.
5. Input the shrinkage computation equation (i.e. shrinkage (%) = MCA × Multiplier) using the appropriate Excel cell references.

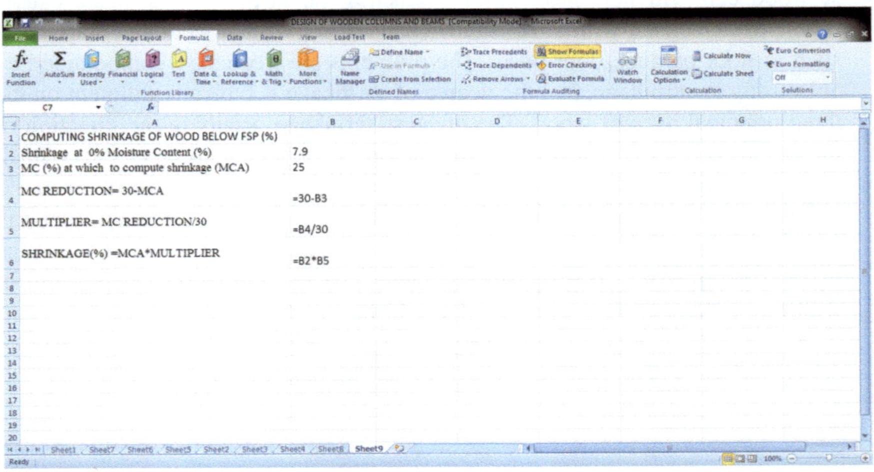

Fig. 4.1 Excel worksheet showing the computations

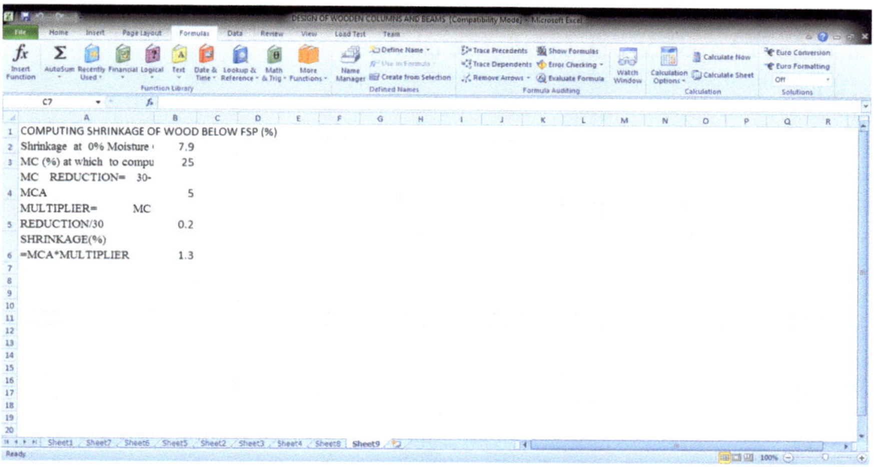

Fig. 4.2 Excel worksheet showing the result

The result of the computation is shown in the Excel worksheet presented in Fig. 4.2.

Example 2 Write a simple FORTRAN program for computing shrinkage of wood at different moisture contents below the fibre saturation point.

Solution

```
! Fortran Program To Compute  Shrinkage of Wood
at Different Moisture Contents Below The FSP
! Knowing The Shrinkage At Oven Dry Moisture Content
WRITE(*,*)'Input shrinkage of the wood at 0% moisture content'
READ(*,*)R
WRITE(*,*)'
Input Moisture content below FSP at which you want to compute shrinkage'
READ (*,*)PN
YM = 30-PN
GN = YM/30
SHR = R*GN
WRITE(*,*)'At a Moisture content of 0.0%'
WRITE(*,*)'The shrinkage of wood is',R,'%'
WRITE(*,*)'At a Moisture content of',PN,'%'
WRITE(*,*)'The shrinkage of wood is',SHR,'%'
END
```

Example 3 Write a simple FORTRAN program for computing actual shrinkage in dimension of wood at a given moisture content.

Solution

```
! FORTRAN Program to Compute Actual Shrinkage in Dimension of Wood
at a Given M.C BFAC = 25.4
WRITE(*,*)'Input shrinkage value of the wood species'
READ(*,*)R
WRITE(*,*)'Input the moisture content'
READ(*,*)CM
WRITE(*,*)'Input original width or thickness of the wood member in inches'
READ (*,*) PN
WM = PN*BFAC
YM =  (CM/3.0)*(1.0/32.0)
SHR = YM*PN
ASHR = PN-SHR
TW =  SHR*BFAC
ATW = WM-TW
WRITE(*,*)'At a Moisture content of',CM, '%'
WRITE(*,*)'and a shrinkage value of',R, '%'
WRITE(*,*)'The width/thickness before shrinkage is',PN, 'inches',
   WRITE(*,*)'or',WM, 'mm'
   WRITE(*,*)'The actual shrinkage is',SHR,'inches'
   WRITE(*,*)'or',TW,'mm'
   WRITE(*,*)'The width/thickness after shrinkage is',ASHR, 'inches',
   WRITE(*,*)'or',ATW, 'mm'
     END
```

Anisotropic shrinkage is mainly responsible for the warping of timber during drying. Whenever the grain does not run parallel to the edges and sides of a board, warping in some form is probable because the differential shrinkage factor will tend to pull the board out of shape. Unequal shrinkage due to unequal drying is also responsible for tissue ruptures such as end splits, surface checks, splits, knots and honeycomb checks. Shrinkage is directly or indirectly responsible for much of the trouble experienced in seasoning wood and wood products. Therefore, the drying method should aim at restraining differential shrinkage as far as possible and at preventing too big a difference in shrinkage between outer and inner parts of the timber.

(ii) Equilibrium Moisture Content

A piece of wood gives off or takes on moisture until the amount of moisture it contains is in equilibrium with that in the surrounding atmosphere. For example, if a piece of wood is exposed to an atmosphere containing moisture in the form of water vapour, it will come after some time to a steady moisture content condition. This equilibrium condition is referred to as equilibrium moisture content (EMC). The EMC depends on the relative humidity, the temperature of the surrounding air and the drying conditions to which the wood has previously been exposed. Drying and shrinkage typically continue until the piece of wood attains EMC, and this should take place before the timber is put into use. The EMC may be as high as 20% in a humid environment as obtained during the rainy season in the tropics and as low as 6% in a hot dry climate as obtained in the peak of the dry season in savanna and desert areas.

In many sub-Saharan African countries, sawn wood products are sold in the 'green' state. The degree of drying that is necessary depends on the use to which a wood product is to be put. In case of woodwork for use in buildings as panelling, doors, furniture, etc., the wood material should be dried to about 12–16% moisture content, depending on the dryness of the locality. Carpentry timbers generally should be dried to below 20% moisture content unless they are employed for rough carpentry purposes in exposed positions where further drying can occur in situ. Timber to be treated with preservatives by pressure treatment should be dried to about 25%.

4.4 Seasoning Methods

There are principally two ways to season wood: air seasoning under natural conditions or subjecting the material to an artificial seasoning process in a chamber (referred to as a kiln), whereby the process is accelerated and the various drying parameters of temperature, humidity and air speed are carefully controlled. In both air- and kiln-drying methods, the moisture is removed by evaporation from the surface of the timber. It has already been said that air-drying depends on wind, sun and rain. The rate of evaporation is partly controlled and therefore depends largely on the prevailing weather conditions. In the dry season, drying may be rapid, and in the rainy season, it may be very slow. It is possible to dry timber down to about 12% in the dry season in humid climates and still further in the dry regions, but 18% is a more usual figure in the rainy season in the humid tropics.

4.4.1 Air Seasoning

Much of the timber air-dried in the tropics is piled out-of-door without any protection beyond the use of some primitive form of roof in some instances. Timber products air-dried in this manner include railway sleepers, poles and planks (see Figs. 4.3 and 4.4). However, it is generally advisable to pile timber in a seasoning shed with a pitched roof having an adequate overhang to serve as a protection against driving rain. In the wetter parts of the tropics, it is generally sufficient to have open-sided seasoning sheds with good deep eaves. In the drier regions,

Fig. 4.3 Posts and poles laid out to dry

Fig. 4.4 Hardwood lumber of pile-stacked for air-drying

however, it is recommended that additional side protection be provided to reduce overrapid drying in very hot, humid, dry periods.

Contrary to the general tendency to assume that air seasoning is a poor substitute for kiln seasoning, it is not. In fact, a combination of the two methods is usual and often desirable. For example, there are some woods, in thick sizes like 75 mm (3 inches) that cannot be successfully kiln-dried from the green state, but have to be air-dried for long periods before final kiln-drying. In both air and kiln seasoning, the moisture is removed by evaporation from the surface of the timber. It has already been said that air seasoning depends on wind, sun and rain. The rate of evaporation is partly controlled and therefore depends largely on the prevailing weather conditions. In the dry season in the tropics, drying may be rapid, while in the rainy season it may be very slow. It is possible to dry timber down to about 12% in the dry season, but 18% is a more usual figure in the wet season. These sidewalls should allow free circulation of air. They should therefore be raised well off the ground and should also leave several feet open under the eaves. These sidewalls may be removed in the wet (rainy) season.

Table 4.1 shows selected tropical hardwood timbers and their drying properties. Easy drying hardwood timbers may be stacked for drying at any time, but the best time to pile most timbers for air-drying is during the cool rainy season so that early drying conditions may be mild. Timber that has been dried during the dry season will hardly dry at all during the following wet season, and hence, it is advantageous to arrange that the drying process be completed by the beginning of the rains.

Degrades associated with poor air-drying include checks, splits, warps, insect and fungal stains and are usually noticeable in the lumber produced. The resultant effects include loss in mechanical strength, reduced durability and general loss of structural integrity, aesthetic appeal and economic value. Adoption of the following practices will help in reducing degrades and attendant losses associated with air-drying:

(a) **At the Drying Site**

- Avoid damp, poorly drained sites which have the tendency of reducing the drying power of the air and creating a situation whereby moisture is picked up from the ground as well as from the timber.
- Keep the drying yard sealed and well drained.
- Ensure that the site is far removed from boundary hedges and buildings which tend to impede airflow.
- Keep the drying yard free of vegetation—grass and weeds—and if possible, cover the soil with ashes or treat it in some other ways likely to prevent any renewed growth of vegetation. This is necessary to encourage air movement (circulation) beneath and around the stacks, thereby ensuring that the timber at the base of the pile dries at same rate as those above and also preventing fungus growth.
- Remove wood waste (sawdust, offcuts) from the yard as this is a fire hazard and provides opportunities for fungi and insects to breed.

(b) **Stacking Procedure**

- Stack the timber outside, preferably under shelter. Some covering materials should be used on top of zinc or aluminium roofs to reduce the amount of radiant heat, which would cause fast drying and degrade.
- Ensure that the stacks are at least 23 cm (9 inches) proud of the ground, preferably resting on very durable members (e.g. *Lophira alata*) spaced at most 4 intervals.
- Align stacks parallel to the prevailing wind direction.

Table 4.1 Seasoning properties of selected tropical hardwood timbers

Hardwood species	Seasoning	Properties	
	Drying rate*	Movement**	Shrinkage***
	R, FR, RS S, VS	L, M, S	L, M, S
Mitragyna ciliata	R	S	M
Terminalia superba	R	S	S
Lovoa trichiloides	FR	N/A	M
Afrormosia elata	RS	S	S
Gossweilerodendron balsamiferum	FR	S	S
Albizia spp.	VS	S	N/A
Alstonia boonei	R	N/A	M
Anogeissus leiocarpus	VS	N/A	M
Antiaris africana	FR	S	S
Afzelia africana	VS	S	S
Distemonanthus benthamianus	R	S	N/A
Berlinia spp.	RS	M	M
Pterocarpus soyauxii	RS	S	S
Canarium schweinfurthii	RS	M	M
Ceiba pentandra	N/A	N/A	N/A
Celtis spp.	FR	M	M
Cordia millennii	N/A	N/A	N/A
Daniellia ogea	N/A	N/A	N/A
Nesogordonia papaverifera	RS	M	M
Diospyros piscatoria	FR	S	M
Piptadeniastrum africanum	S	M	M
Isoberlina doka	VS	M	M
Diospyros mespiliformis	FR	S	M
Lophira alata	VS	M	M
Ricinodendron heudelotii	R	N/A	S
Erythrophloeum guineense	S	S	N/A
Combretodendron macrocarpum	S	L	L
Entandrophragma angolense	FR	S	M
Guarea spp.	FR	S	M
Holoptelia grandis	N/A	N/A	N/A
Terminalia ivorensis	R	S	S
Pycnanthus angolensis	N/A	N/A	N/A
Milicia excelsa	FR	S	S
Oxystigma oxyphyllum	N/A	N/A	N/A
Khaya grandifoliola	FR	S	M
Khaya senegalensis	FR	S	M
Khaya ivorensis	FR	S	M
Mimusops djave	FR	S	M

(continued)

Table 4.1 (continued)

Hardwood species	Seasoning	Properties	
	Drying rate*	Movement**	Shrinkage***
	R, FR, RS S, VS	L, M, S	L, M, S
Mansonia altissima	FR	S	M
Triplochiton scleroxylon	FR	S	S
Scottellia coriacea	FR	M	M
Daniellia ogea	FR	M	M
Cylicodiscus gabunessis	S	N/A	M
Brachystegia spp.	S	M	M
Entandrophragma candollei	RS	M	M/L
Nauclea diderrichii	FR	S	S
Pterocarpus erinaceus	RS	S	S
Pterygota macrocarpa	RS	N/A	N/A
Entandrophragma cylindricum	FR	N/A	M
Sterculia rhiopetala	VS	L	L
Sterculia oblonga	S	M	M/L
Tectona grandis			
Entandrophragma utile	FR	M	M

Source Lucas (1983), NCP (2005)
Legend: ***Drying Rate**: *R* Rapid, *FR* Fairly rapid, *RS* Rather slow, *S* Slow, *VS* Very slow
****Movement**: *L* Large, *M* Medium, *S* Small
*****Shrinkage**: *L* Large, *M* Medium, *S* Small
Note The quantitative values of these parameters are not readily available

- Place stickers of uniform dimensions (typically 15–25 mm thick, 20–75 mm wide) produced preferably from the heartwood of a well-seasoned lumber between the layers of timber in the stack.
- Spray the stack once a week with fungal or insect-destroying chemicals. Spraying needs not be done by using expensive machines as it could be done manually.

4.4.1.1 Advantages of Air Seasoning

The following are some of the advantages of air seasoning:

- The method is simple and cheap.
- The process allows the timber to dry slowly, thereby reducing the possibilities of encountering timber degrades, if the process is well-managed.
- Air-drying, if properly conducted, is suitable for many impermeable or collapse-prone species, large size timber members and items for exterior use which do not require a low final moisture content.

- Kiln-drying of green, impermeable hardwoods takes too long and is uneconomical. Instead, if faster or more controlled drying is required, the timber should be air-dried first to around the fibre saturation point before being kiln-dried.
- Where the climate is humid, degrade in air-drying of non-perishable timber is usually small, provided the green timber is put out to dry during the rainy season.

4.4.1.2 Disadvantages of Air Seasoning

The major disadvantages of air seasoning include the following:

- It is often more expensive than it seems. This is because land is required and large quantities of stock have to be held to justify the investment. With hardwoods, the volume of stock may equal the annual turnover.
- There is little control over the drying elements (wind and wind direction, temperature and sunshine, humidity and rain) and therefore the entire process.
- The drying time is very variable depending on the location, season and the species.

4.4.2 Artificial Seasoning (Kiln-Drying)

Artificial seasoning can be either moderate or rapid depending on the temperature of the air injected into the chamber where the timber is piled and the rate at which the air is circulated and extracted from the chamber. It is a much faster method of drying. Many hardwoods can be dried in three weeks or less, although some must be partially air-dried beforehand. Therefore, orders can be filled at short notice. However, this method is expensive and can only be applied on small quantities of timber. Timber can be artificially seasoned from the green condition, but often hot air seasoning is used only at a later stage after most of the moisture has been removed with air seasoning. In many parts of tropical Africa, however, conventional kiln dryers are still not common, their use being limited to a few private wood processing companies. The use of solar kilns as an efficient and economic means of drying wood in tropical areas where sunshine is available throughout the year is a potential sustainable alternative. This is simply because solar kilns are relatively cheaper than conventional kilns to install, operate and maintain.

Smoke seasoning is a moderate process and involves placing the timber over a bonfire. It can take a month or two depending on the size and type of wood being seasoned. This method is considered to be both a seasoning and a treating method for timber. Presumably, it protects the timber against pest attacks and increases durability. However, it is not very reliable and can lead to splitting of the timber because of lack of control of the heat from the bonfire.

Seasoned timber should be protected from moisture on the building site. Close piling and covering with tarpaulins delays the absorption of atmospheric moisture, particularly in the interior of the pile.

4.4.3 *Milling and Seasoning Defects*

There are often residual stresses in a log delivered to the mill causing some deformation in the sawn board which may be emphasized during the drying process and shown as:

- **Curvature**—bowing and spring over the board length as shown below

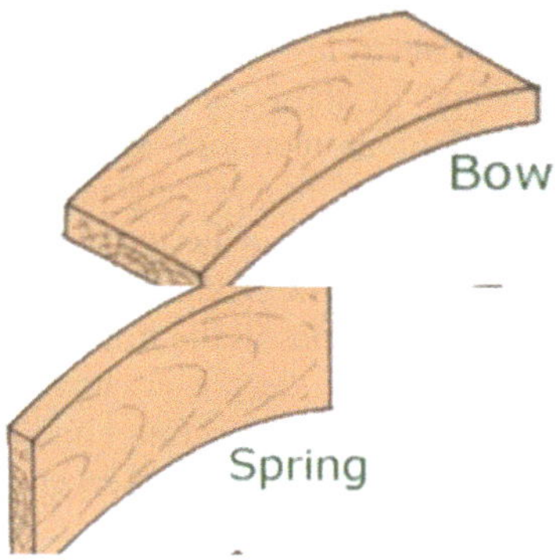

- **Warping**—cupping in board cross-section or twisting over the board length as shown below:

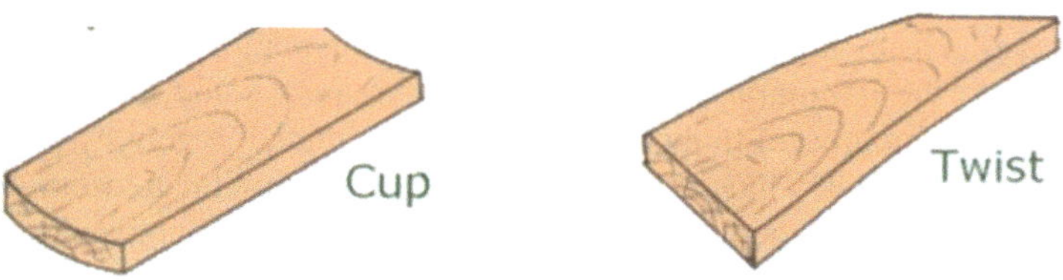

- **Splits** or longitudinal cracks that go right through the thickness of the board.
- **Seasoning checks** due to uneven shrinkage and seen as small fibre separations on the surface of a board, or as a longitudinal crack perpendicular to the growth rings. Such checks could also appear in seasoned wood that is exposed to large humidity changes even with a water vapour inhibiting barrier.

4.5 Hardwood Timber Preservation

When wood that is not naturally decay-resistant is required in a wet application, or where it may be at risk for insect attack, it is necessary to specify preservative-treated wood. This is lumber that has been chemically treated to make it unattractive to fungi and other pests. In the same way galvanized steel may be specified where it would be at risk of rusting, treated wood may be specified in a setting conducive to decay. Also like galvanized steel, treated wood has a shell of protected material surrounding an unprotected core. As long as this shell stays intact, the core will be protected.

Wood does not deteriorate just because it gets wet. When wood breaks down, it is because an organism is eating it as food. Preservatives work by making the food source inedible to these organisms. The principal agencies that cause wood deterioration or destruction are weather, fire, fungi, insects and marine borers. Adequate protection from the influence of these agents may prolong the life of timber indefinitely. Wood preservation can therefore be broadly defined as the protection of wood from fire, chemical agents (e.g. strong acids and alkalis when stored in wooden vats), mechanical wear and weathering, as well as biological attack by agents such as fungi and insects. The act of fire-proofing timbers has not been seriously pursued yet in many countries in the tropics, perhaps due to the relatively high cost of fire retardants. Most of the preservative treatments are performed to protect timber products against attacks from fungi and the major economic pests—mainly beetles and termites. Hence, wood preservation against fungi and insect attack will be the major focus here.

Different species of hardwood timber vary considerably in their resistance to one or more of the destructive agents. While some tropical hardwood timbers are very resistant to insect and fungal damage and may have a comparatively long life even when used in situations where they are very liable to decay, majority of them get severely damaged in such conditions unless they are thoroughly treated with wood preservatives. Even though used in conditions unfavourable to fungal decay or termite attack, the non-durable timbers generally require some form of treatment. The durability of untreated timber is, however, determined largely by the presence or absence of sapwood, which is the part that is most susceptible to decay. Sapwood is therefore classified as perishable (<5 years) or at best non-durable (5–10 years), while heartwood is classified in terms of its durability as perishable or durable. The classifications are usually arrived at on the basis of ground contact field (graveyard) tests.

Unlike mould and staining fungi, wood-destroying fungi seriously reduce strength by metabolizing the cellulose fraction of wood that gives wood its strength. Early stages of decay are virtually impossible to detect. For example, brown-rot fungi may reduce mechanical properties in excess of 10% before a measurable weight loss is observed and before decay is visible. When weight loss reaches 5–10%, mechanical properties are reduced from 20 to 80%. Decay has the greatest effect on toughness, impact bending and work to maximum load in bending, the least effect on shear and hardness and an intermediate effect on other properties. No method is available yet for estimating the amount of reduction in strength from the appearance of decayed wood. Thus, when strength is important, adequate measures should be taken to: (a) prevent decay before it occurs, (b) control incipient decay by remedial measures or (c) discard and replace any wood member in which decay is evident or believed to exist in a critical section. An exception may be pieces in which decay occurs in a knot but does not extend into the surrounding wood. Decay can be prevented from starting or progressing if wood is kept dry (below 20% moisture content). The major advantage of chemical preservative treatment is that it increases perishable life of timber in the order of up to 40 years.

There are basically two conditions of wood exposure that require preservative treatment. These are as follows:

Before utilization: Temporary preservative treatment may be required when wood is stored in damp areas. Also, between time of conversion and time of final conditioning, lumber and veneers may require preservative treatment to resist attack by wood-staining fungi and wood-boring insects.
During Utilization: There are basically five service conditions that call for preservative treatment of wood. These are:

- **Service Condition I**: Wood in permanent contact with the ground or close to a persistent humidity source, e.g. timbers, wood paving blocks, piling in freshwater, fencing post, railway sleepers, telegraph and transmission poles.
- **Service Condition II**: Wood not in contact with the ground but subjected to long period of rehumidification, e.g. greenhouse, heavy duty flooring in trucks and box cars, cooling tower.
- **Service Condition III**: Wood not in contact with the ground but subjected rehumidification by rain, e.g. exterior joinery and structural formwork.
- **Service Condition IV**: Wood not in contact with the ground and not exposed to weather, e.g. interior joinery, furniture and carving.
- **Service Condition V**: Wood used in sea or brackish water, e.g. marine constructions and harbour works.

The choices involved in timber preservation include choice of timber to be treated, choice of the preservative to be used in conjunction with that timber and the quantity of preservative needed, and the choice of treatment process since not all treatment combinations will be acceptable for a particular end use. A major challenge in the treatment process is to get the toxic chemicals sufficiently deep into the wood for long-term protection. The selection of treatment is determined by both technical and economic considerations. An ideal wood preservative should possess the following basic features:

- Ability to penetrate wood,
- Easy application,
- Chemical stability at all times, not being removed by water,
- Permanent fixity into wood,

- Toxicity to wood destroyers—fungi and insects,
- Not harmful to wood, metal and/or man,
- Safe and economical to use,
- Non-flammability,
- Colourless or good paintability.

The efficacy of any wood preservative treatment can be determined based on the following three parameters, considered simultaneously: *penetration*, *distribution* and *retention* in wood. The net retention of preservatives by treated wood is therefore not a sufficient indication of the efficacy of treatment since the preservative can be concentrated in certain areas of the wood, leaving variations in depth of penetration. Even distribution is desired so that no part of the wood is less well-protected than others.

There are several chemical formulations and treatment methods already devised for preserving timber. A full discussion of these chemical products and application methods is beyond the scope of this book. We shall, however, discuss briefly some of the wood preservatives currently in common use which can be broadly divided into three groups, based on the type of solvent used, i.e. oil-based types, waterborne types and organic solvent types. Oil-based preservatives are the most common and efficient preservatives in general use for treating structural timber, building timbers exposed to high decay hazard, timber either in ground contact or exposed to high decay and termite hazards, fence posts, railway sleepers and ties, marine structures, etc. Examples include coal-tar oil, creosote and solignum. In addition to being effective fungicides, these preservatives are highly toxic to timber insects and marine borers. They are satisfactory as regards penetration and permanence since they are relatively insoluble in water. They are also relatively cheap. Using oil-based preservatives however has the following disadvantages:

- They have poor staining characteristics and usually cannot be painted over satisfactorily. Their dark colour and oily nature make them unsuitable for use on surfaces that may subsequently require painting or polishing.
- Their pungent odour, though in time diminishing, precludes their use for certain purposes and may not be permitted in certain areas in a structure, e.g. kitchens or even inside dwelling.
- They have a relatively high initial fire hazard (after a few weeks, the more flammable fractions of the oil volatize from the surface).

The common waterborne preservatives include water-soluble salts, such as zinc chloride, zinc–chrome–arsenic, copper–chrome–arsenate, copper sulphate and arsenic. These preservatives are relatively cheap, odourless, clean to use and can be painted over when the treated timber becomes dry. They also have the advantage of being supplied in powder form. Zinc chloride, however, is corrosive to metals. Steel and wire nails may be badly corroded under very humid conditions, especially when used in unseasoned treated wood. In normal circumstances, corrosion is not serious and may be discounted. Zinc chloride is also very soluble in water. Hence, it has the tendency of leaching out of the timber exposed to wet conditions. It is generally unsuitable for outdoor use, but it is a reliable preservative for interior use. Two to five per cent solutions are generally used, it being toxic to fungi at 1 or 2% concentrations.

Copper sulphate is now rarely used alone, but it is an ingredient of a number of proprietary preservatives. It is rather more toxic to fungi than zinc chloride and, like it, is readily leached from wood. It is also corrosive to iron and steel, and it cannot be used in ordinary treating plants. Arsenic salts are also rarely used alone. They are frequent constituents of patented wood preservatives. They are highly toxic to insects but generally are less toxic to fungi.

Organic solvent types of preservatives include chemical compounds such as naphthenates (i.e. copper, iron and zinc naphthenates) and chlorinated phenols which are dissolved in volatile oils or spirits. They are used for both immunization pre-treatment and eradication of existing attack. They are rapid drying and clean to handle. They penetrate wood comparatively well. Good penetration is achievable with light solvents such as kerosene. They are, however, relatively more expensive. In the last few decades, concerns have been raised that preservatives may leach from the wood into surrounding soil, resulting in environmental pollution. Hence, a lot of research is currently going on regarding the development of environmentally benign and non-metallic wood preservatives.

The various methods of applying wood preservatives in general use across the tropics include the following:

Brushing or Spraying: This is a simple and inexpensive method that is moderately effective if the wood is not exposed to severe attack by fungi and borers. It affords only very limited protection against termites, and in a damp situation, the preservative is likely to be washed off before long. It is not usually used for timber in contact with the ground or one that is exposed to the weather since only a very light surface penetration can be obtained and subsequent abrasion may expose to untreated timber. When treated timber is used in an exposed condition, brush treatments are repeated at regular intervals to maintain the effectiveness of preservative used.

Dipping: This involves dipping the timber in a tank filled with the preservative. A simple wooden trough is generally adequate for dipping, but more elaborate plant is used when large quantities of timber have to be treated. It is somewhat more effective than brushing as it usually ensures that all cracks and openings in the timber are treated, and a better penetration is usually obtained. Time of application may vary considerably, depending on the condition of the timber (i.e. whether green or seasoned) and the type of preservative employed (i.e. whether soluble or insoluble in water). Water-soluble salts tend to be more effective in treating green timber, while oil soluble preservatives are the most effective in treating seasoned timber.

Steeping: This involves soaking the timber in a preservative for prolonged period of time, up to several weeks for improved penetration which is usually not guaranteed. The method is also time-consuming. A short period of dip is only a little more effective than brushing or spraying, but prolonged steeping is effective with water-soluble salts as diffusion goes on. Most of the preservative is absorbed in the first few days but continues to be absorbed for much longer periods. Hardwood timbers that are readily penetrable, e.g. *Antiaris* and *Alstonia* species, as well as the sapwood of many other species, may be given good protection by this means.

Dip Diffusion: This is an improved method of dipping that involves momentary dipping (or spraying) of green timber in a concentrated solution of the preservative followed by block stacking, for a period of more than three weeks, under plastic sheets or in a sealed room to prevent drying, while the preservative diffuses sufficiently into the timber.

Hot and Cold Bath Process (Open Tank): This is a method by which seasoned timber is immersed in a tank of cold preservative which is then heated to about 90 °C and maintained for a fixed period (usually a few hours). The heating is thereafter discontinued, and the hot preservative is then allowed to cool. During heating, the hot liquid heats the timber and causes the air in the timber to expand and a certain amount is expelled. On cooling, the remaining air in the wood contracts and forms a partial vacuum and the preservative is sucked in. Some control of absorption and penetration can be obtained by manipulating the periods of hot and cold dip and temperature drop. The cooling can be accelerated by means of water-cooling coils, or the timber may be quickly removed to a second tank containing cold preservative, or again, the hot preservative may be run off to another tank and replaced with a cold one.

The periods of heating and cooling depend on the permeability of the timber and the efficiency of heating and cooling. Timbers that are very resistant to penetration will not absorb much preservative by this method, but the moderately impervious timbers can be treated fairly adequately by 4 h heating and 20 h cooling. Periods of heating and cooling should be increased if absorption is inadequate. Timbers that are moderately permeable will not require as long treatment as this, and those that are readily permeable may be treated in a much shorter time.

Vacuum Processes: These less common processes are usually carried out in a special plant, and they involve the use of a pressure pump, a vacuum pump, a cylinder capable of working at pressures up to 1379 kPa, a mixing tank and storage tanks for the preservative. The *full-cell process* involves impregnating the cell and filling the cellular spaces in the structure of the timber with preservative solution and subsequently allowing the timber to dry out, thereby removing the water and leaving the preservative within the cell structure. The *Lowry (empty-cell) process* is usually employed for large-sized impermeable timbers when limited time is available for drying.

For preservatives to work effectively, it is desirable that the wood is permeable so that the preservatives can penetrate readily. A wide permeable sapwood band is also preferred to the heartwood since the durability of treated sapwood can be considered greater than that of untreated heartwood, and the heartwood of many hardwood timbers is either less permeable or refractive (i.e. resistant to penetration by preservatives), making it much harder to treat with preservatives. Tropical hardwood timbers vary a great deal in their ability to absorb preservatives. Some are practically impermeable even under pressure treatment, and so it is not practicable

Table 4.2 Durability and resistance to impregnation of selected tropical hardwood timbers

Timber species	Natural durability	Resistance to impregnation
Afrormosia elata	Very durable	Extreme
Afzelia africana	Very durable	Extreme
Albizia spp.	Very durable	Extreme
Alstonia boonei	Perishable	Permeable
Anogeissus leiocarpus	Durable	Resistant
Antiaris africana	Perishable	Permeable
Berlinia spp.	Durable	Extreme
Brachystegia spp.	Non-durable	Extreme
Canarium schweinfurthii	Perishable	Permeable
Ceiba pentandra	Perishable	Permeable
Celtis spp.	Perishable	Permeable
Cordia millennii	Moderate	Resistant
Cylicodiscus gabunessis	Very durable	Extreme
Daniellia ogea	Perishable	Very resistant
Distemonanthus benthamianus	Very durable	Resistant
Entandrophragma angolense	Moderate	Extreme
Entandrophragma utile	Durable	Extreme
Gossweilerodendron balsamiferum	Very durable	Very resistant
Holoptela grandis	Moderate	Moderate
Khaya grandifoliola	Durable	Very resistant
Khaya ivorensis	Durable	Resistant
Khaya senegalensis	Durable	Very resistant
Lophira alata	Very durable	Extreme
Lovoa trichiloides	Moderate	Resistant
Mansonia altissima	Very durable	Very resistant
Milicia excelsa	Very durable	Resistant
Mimusops djave	Very durable	Resistant
Mitragyna ciliata	Non-durable	Permeable
Nauclea diderrichii	Very durable	Resistant
Nesogordonia papaverifera	Moderate	Resistant
Scottellia coriacea	Durable	Permeable
Sterculia oblonga	Non-durable	Extreme
Sterculia rhiopetala	Moderate	Extreme
Tectona grandis	Durable	–

(continued)

Table 4.2 (continued)

Timber species	Natural durability	Resistance to impregnation
Terminalia ivorensis	Durable	Extreme
Terminalia superba	Non-durable	Permeable
Tripochyton scleroxylon	Non-durable	Moderate

Source NCP (2005)
Legend
Extremely resistant: Timbers in which impregnation by prolonged open tank treatment is insignificant and therefore impracticable
Very resistant: Timbers which may be given moderate protection by 4 h/20 h open tank treatment. Lateral penetration is about 3–1.5 mm and end penetration 1.5–13 mm. Some of these timbers show no penetration beyond the limits indicated, some show a very patchy penetration to some depth, while some allow good penetration along vessels. Treatment by a pressure process is therefore advisable
Resistant: Timbers that allow reasonably good penetration by the open tank process. A full 4 h/20 h treatment is generally necessary. Lateral penetration is over 1.5 mm and may be considerably more, but absorption is not very heavy. End penetration is usually deep
Moderately Resistant: Timbers that allow good penetration by the open tank process. A heavy absorption and deep penetration are obtained by 4 h/20 h treatment. Most of these timbers would be given moderately good protection by prolonged cold steeping
Permeable: Timbers that can absorb a very large quantity of preservatives and that can generally be given almost complete impregnation by open tank treatment. They may be satisfactorily treated by cold steeping. No elaborate impregnation plant is required for treating them

to treat them. Others are resistant to penetration, and only limited protection can be given by treatment, impregnation being either superficial or patchy. The advisability of treating such timbers depends on the use to which they will be put. At the other end of the scale are timbers that absorb preservatives readily and which may be adequately treated by steeping in cold preservatives. The durability and resistance to impregnation of selected tropical hardwood timbers are presented in Table 4.2.

Practice Questions

1. Discuss the various reasons for human intervention timber drying.
2. Assuming that the total tangential shrinkage of a tropical hardwood species dried to 0%, moisture content is 10.5%. Use Microsoft Excel to compute its tangential shrinkage if the moisture content of the environment is 27%.
3. Discuss the advantages and disadvantages of air-drying timber.
4. Describe three methods of applying timber preservatives.

References

Adefisan OO, Ajala OO (2016) Emerging trends in wood seasoning and preservation. In: Olorunnisola AO (ed) Renewable natural resources engineering: essays in honour of Canon Prof. E. Babajide Lucas. Zenith Book House Publishers, Ibadan, pp 181–198

Dinwoodie JM (1981) Timber, its nature and behaviour, international student edition. Van Nostrand Reinhold, New York

Forest Products Laboratory (2010) Wood handbook—wood as an engineering material, general technical report FPL-GTR-190. The U.S. Department of Agriculture, Forest Service, Forest Products Laboratory. Madison, WI

Lucas EB (1983) Factors preventing wider commercialization of Nigerian trees. For Prod J 33 (5):64–68

NCP (2005) Nigerian code of practice for structural timber design, published by the Standards Organisation of Nigeria

Olorunnisola AO, Omoniyi TE (2016) Wood conversion and processing in Nigeria: a status report. In: Olorunnisola AO (ed) Renewable natural resources engineering: essays in honour of Canon Prof. E. Babajide Lucas. Zenith Book House Publishers, Ibadan, pp 101–121

Ramussen EF (1960) Dry kiln operator's manual, agricultural handbook No. 188. The U.S. Department of Agriculture, Forest Service, Forest Products Laboratory. Madison, WI, USA

Walker JCF, Butterfield BG, Langrish TAG, Harris JM, Uprichard JM (1993) Primary wood processing principles and practice. Chapman and Hall Publishers, London

Chapter 5
Uses of Wood and Wood Products in Construction

5.1 Emergence of Different Wood Products as Structural Materials

Hardwood timbers and other wood products have remained major structural materials for building construction worldwide. Since ancient times, wood in different forms have been used for myriad structural purposes. With advancements in science and technology over the years, involving the development of sawing and other wood processing machines and a greater understanding of adhesive science and technology, newer wood products for structural utilization have continued to be introduced into the global market. Factors that favour the use of wood products for structural purposes include their renewable nature, availability in various sizes, shapes and colours, affordability, relatively high fatigue resistance and specific strength, ease of joining, durability and aesthetic appeal. Besides, unserviceable wooden building components are recyclable either for their structural properties, e.g. reused permanently as framing or temporarily as form work, or for their heat content as fuel.

On a volume basis, wood products comprise almost 80% of all materials used in the construction of over 1.9×10^6 housing units erected in the USA in 2004 alone. It has also been estimated that over 7×10^9 m^3 of lumber has gone into structures in the USA over the last century. In many other parts of the world, wood products of various types continue to be used in framing, flooring, roofing, siding and even foundation work in single family residences, apartment, commercial and industrial buildings, farm dwellings and service buildings. In Nigeria, for example, the building industry alone consumes about 80% of the country's estimated 20 million cubic meters of annual lumber production. The sawmilling industry, which thrives mainly in the southern part of the country, currently supplies lumber largely for the local market. Also, the numerous small-scale wooden furniture establishments in

A.O. Olorunnisola, *Design of Structural Elements with Tropical Hardwoods*,
DOI 10.1007/978-3-319-65343-3_5

the country consume over 200,000 cubic metres of lumber annually. Figure 5.1 shows the different generic wood products and the sequence in which they emerged as structural materials.

5.2 Unprocessed and Semi-processed Solid Wood

In the ancient past, there were no items of equipment to convert solid wood prior to its structural utilization. Therefore, solid wood of different sizes was being used in the raw unprocessed form. Even with the development of wood conversion equipment of various kinds, solid wood has remained a major structural material in many parts of the globe. The different types in use include the following:

(i) **Logs**: A log is a section of the trunk of a tree in suitable length and diameter for sawing into commercial lumber, but which has not been sawn. Logs were the first set of traditional wood products used for structural purposes before the advent of the saw which made wood conversion possible. In many parts of the tropics, logs were and are still used in building log cabins (Fig. 5.1), barns, bridges, etc., and as columns, beams and rafters in building construction (Figs. 5.2 and 5.3).

(ii) **Poles**: These are straight pieces of 5 m or more in length taken from trunks of trees and used principally to support telephone, electrical transmission lines and for scaffolding, after due treatment with preservatives. They are also used as studs, columns (Fig. 5.4), beams, wall plates, rafters and purlins (Fig. 5.5) largely in farm structures and rural residential buildings, and for scaffolding in multi-storey building construction sites.

(iii) **Posts**: These are round, hewn, squared or split wood, usually less than 3 m in length, but possibly up to 5 m, used principally for scaffolding and rafting (Fig. 5.6) for farmhouses, sheds, livestock buildings, storage structures, beams for drying platforms and generally for fencing (Fig. 5.7).

On the positive side, the use of wood in the round, unconverted forms as logs, poles and posts has many advantages. These include the following:

- Availability for use in rural settings that do not have facilities for log conversion.
- Affordability since the processing cost is almost nil.

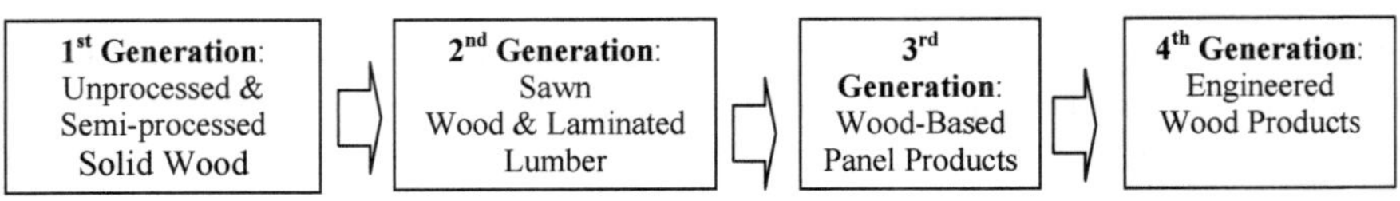

Fig. 5.1 Sequence of emergence of different wood products

Fig. 5.2 A log cabin

Fig. 5.3 Logs used as columns in a rural residential building

- Natural durability (in many instances) since the heartwood would still be intact.
- Greater strength since the wood fibres have not been separated through wood conversion.
- Relatively high fire resistance compared to converted lumber, given the larger size involved.
- Less wood waste generation compared to sawn wood.
- Aesthetic appeal.

On the negative side, the use of wood in the round, unconverted forms as logs, poles and posts is disadvantageous in some ways:

Fig. 5.4 Small diameter logs used as columns, beams and rafters

Fig. 5.5 Wooden poles used as wall plates, rafters and purlins

- Wide variability in size and shape
- Difficulties associated with handling and joining, particularly large diameter logs.
- Difficulties in drying thoroughly in good time which may result in dimensional instability.
- Difficulties in administering preservative treatment.

Fig. 5.6 Wooden posts and poles used as rafters

Fig. 5.7 Wooden post used for fencing

- Retention of natural defects.
- Durability issues since the perishable sapwood is not separated from the more durable heartwood.
- Age-related strength issues—the tree has to be old enough (≥ 20 years) to produce strong timber.
- Non-suitability for certain structural applications, e.g. truss fabrication.

5.3 Sawn Wood

With the advent of wood conversion machinery, it became possible to process wood by sawing, resawing, passing it lengthwise through a standard planing machine, cross-cutting to length and/or matching it into sawn wood (also commonly referred to either as lumber or as sawn timber) prior to structural utilization. With the development of the saw blade and various types of sawing machines, sawn wood emerged and has remained the most common wood product in many countries in sub-Saharan Africa. It is used largely in building construction as beams (Fig. 5.8), furniture making, joinery, crates making, etc.

On the positive side, sawn wood is:

- Relatively affordable,
- Easy to handle, machine, join and maintain,
- Very light (it has a relatively high strength/weight ratio),
- Aesthetically acceptable,
- Naturally durable, depending on the species and form of utilization.

On the negative side, sawn wood:

- Is susceptible to bio-deterioration if not properly handled and/or treated with appropriate wood preservatives.
- Is susceptible to weathering agents if not adequately protected.
- Is susceptible to manufacturing defects if not properly handled, and considerable wood waste may accompany the conversion and processing, e.g. sawdust, slabs, mis-manufactured lumber.
- Is usually not readily available and/or affordable to rural dwellers since additional costs are associated with log conversion and processing.
- Has relatively lower fire resistance, especially the 25-mm-thick planks.

Fig. 5.8 Lumber used as beams

The terminology in the wood industry that is applied to the dimensions of a piece of lumber differs from the terminology normally used in structural calculations. While the industry refers to the thickness and width of a piece of lumber, design calculations usually refer to the width and depth of member. The width is parallel to the neutral axis of the cross-section, while the depth is perpendicular.

The different classes of sawn wood recognized in the global wood products industry include the following:

- **Rough Lumber**: This is lumber that has not been dressed (surfaced) but has been sawn, edged and trimmed at least to the extent of showing sawn marks in the wood on the 4 longitudinal surfaces of each piece for its overall length.
- **Dressed (surfaced) Lumber**: This is lumber that has been dressed by a planning machine for the purpose of attaining smoothness of surface and uniformity of size on: one side (S1S), two sides (S2S), one edge (S1E), two edges (S2E), or a combination of sides and edges (S1S1E, S1S2E, S2 SIE, S4S). In the advanced countries, most structural lumber is dressed lumber, surfaced to the standard net sizes which are less than the nominal (stated) size. Typical lumber will be S4S (surfaced four sides), but other finishes can be obtained, e.g. S2S1E, i.e. surfaced two sides and one edge. Dressed lumber is used in many structural applications, but large timbers are commonly rough sawn to dimensions that are close to the standard net sizes. The cross-sectional dimensions of rough-sawn lumber are approximately 3.2 mm (1/8 in.) larger than the standard dressed size.
  - **Worked Lumber**: This is lumber that has, in addition to being dressed, matched, ship-lapped or patterned.
  - **Structural Joists and Planks**: These are pieces of rectangular cross-section 50–10 mm (2–4 inches) in least dimension, graded primarily for strength in bending edgewise or flatwise but also frequently used where tensile or compressive strength is important. Lumber 50 mm (2 inches) in nominal thickness is often placed in grades separate from the thicker joists and planks (Fig. 5.9).
  - **Beams and Stringers**: These are pieces of rectangular cross-section 5″ × 8″ (nominal dimensions) and larger, graded for strength in bending when loaded on the narrow face.

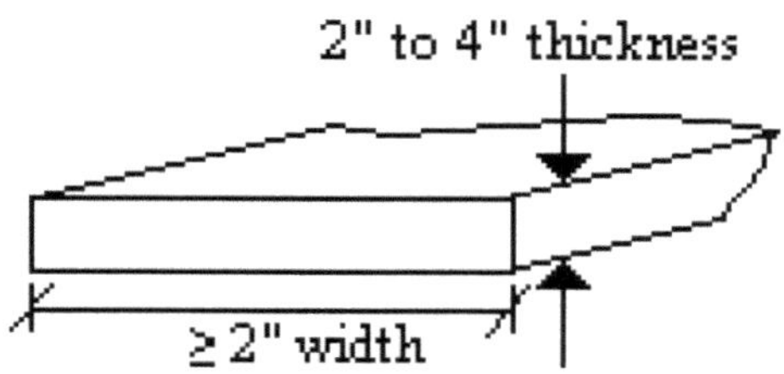

Fig. 5.9 Nominal dimensions of joists and planks

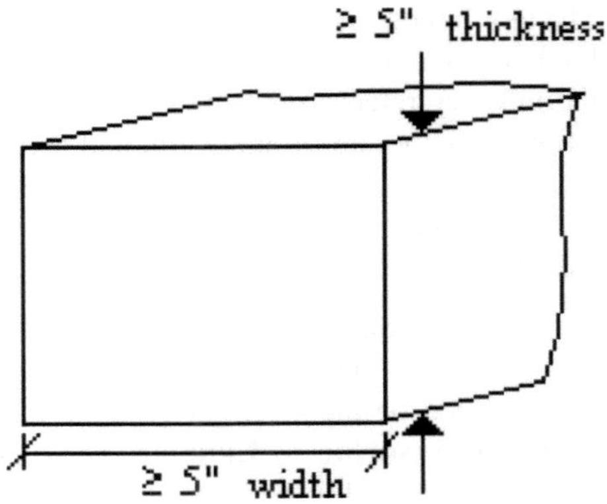

Fig. 5.10 Nominal dimensions of posts and timbers

- **Posts and Timbers**: These are pieces of square or nearly square cross-section 125 × 125 mm (5″ × 5″) nominal dimensions and larger, graded primarily for use as posts or columns (Fig. 5.10).

  In general, posts and timbers are square-like, while beams and stringers are rectangular in shape.

  Structural Boards: These are pieces of lumber that are less than 50 mm (2 inches) in nominal in thickness and of any width, graded primarily for use where the principal stresses are axial in compression or tension (Fig. 5.11).

Lumber sizes commonly produced in Nigerian sawmills are shown in Table 5.1. The 'x' marks indicate the usual nominal lumber thickness/width combinations available. The average length of logs converted is 3.6 m. Basic lengths of timber in eastern and southeastern Africa are between 1.8 metres and 6.3 metres, although pieces longer than about 5.1 metres are scarce and costly. Even though timber is normally sold by length, the price may be calculated per cubic metre when sold in large quantities.

However, in advanced countries, most structural lumber is dressed lumber, surfaced to the standard net sizes which is less than the nominal (stated) size. Typical lumber will be S4S (surfaced four sides), but other finishes can be obtained, e.g. S2S1E, i.e. surfaced two sides and one edge. Dressed lumber is used in many structural applications, but large timbers are commonly rough sawn to dimensions that are close to the standard net sizes. The cross-sectional dimensions of rough-sawn lumber are approximately 3.2 mm (1/8 in.) larger than the standard dressed size.

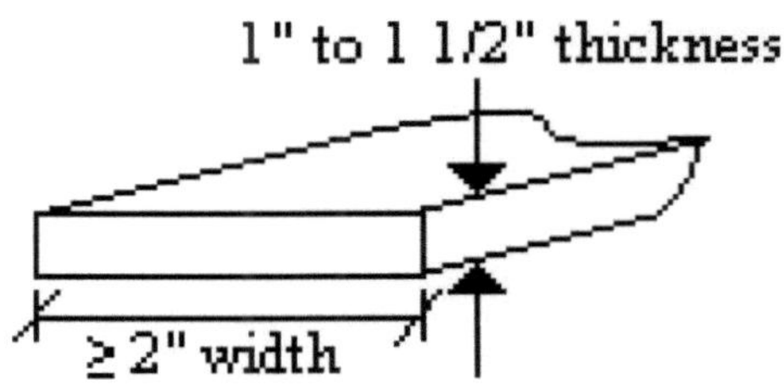

Fig. 5.11 Nominal dimensions of structural boards

Table 5.1 Lumber sizes commonly produced in Nigerian sawmills

Thickness (mm)				Width (mm)					
	75	100	125	150	175	200	225	250	300
16	x	x	x	x					
19	x	x	x	x					
22	x	x	x	x					
25	x	x	x	x	x	x	x	x	x
32	x	x	x	x	x	x	x	x	x
38	x	x	x	x	x	x	x	x	x
44	x	x	x	x	x	x	x	x	x
50	x	x	x	x	x	x	x	x	x
63		x	x	x	x	x	x		
75		x	x	x	x	x	x	x	x
100		x		x		x		x	x
150				x		x		x	x
200						x			
250								x	
300									x

Source NCP (2005)

Table 5.2 Green basic stresses for groups of Nigeria-grown hardwood species at moisture content >18%

Strength group	Bending and tension parallel to grain (N/mm²)	Compression parallel to grain (N/mm²)	Shear parallel to grain (N/mm²)	Compression perpendicular to grain (N/mm²)	Mean value of modulus of elasticity (N/mm²)
N_1	28.0	22.4	3.55	5.00	13,200
N_2	22.4	18.0	2.80	4.00	11,200
N_3	18.0	14.0	2.24	3.15	9500
N_4	14.0	11.2	1.80	2.50	8000
N_5	11.2	9.0	1.40	2.00	6700
N_6	9.0	7.1	1.12	1.60	5600
N_7	7.1	5.6	0.90	1.25	4150

Source NCP (2005)

Lumber grades that provide specific information in structural design are most useful. Nigerian timber species have been divided into seven groups on the basis of strength, i.e. N_1, N_2, … N_7. The strongest timbers fall into N_1 group and the weakest into N_7 group. The strength properties for which basic stress values are given are static bending and tension parallel to the grain, shear and compression parallel to the grain, compression perpendicular to the grain and modulus of elasticity as shown in Tables 5.2 and 5.3. In contrast, the Indian code of practice for the

Table 5.3 Dry basic stresses for groups of Nigeria-grown hardwood species at moisture content $\leq 18\%$

Strength group	Bending and tension parallel to grain (N/mm^2)	Compression parallel to grain (N/mm^2)	Shear parallel to grain (N/mm^2)	Compression perpendicular to grain (N/mm^2)	Mean value of modulus of elasticity (N/mm^2)
N_1	35.5	28.0	4.50	6.30	15,000
N_2	28.0	22.4	3.55	5.00	12,500
N_3	22.4	18.0	2.80	4.00	10,600
N_4	18.0	14.0	2.24	3.15	9000
N_5	14.0	11.2	1.80	2.50	7500
N_6	11.2	9.0	1.40	2.00	6300
N_7	9.0	7.1	1.12	1.60	5300

Source NCP (2005)

design of structural timber in building (1995) classfied species of timber recommended for constructional purposes into three groups, based on their strength properties, namely modulus of elasticity (E) and extreme fibre stress in bending and tension (f_b). Placed in Group A are species with E above 12.6 $x10^3$ N/mm^2, f_b above 18 N/mm^2; placed in Group B are species with E above 9.8 $x10^3$ N/mm^2 and up to 12.6 $x10^3$ N/mm2, f_b above 12.0 N/mm^2 and up to 18 N/mm^2; while the species placed in Group C are those with E above 5.6 $x10^3$ N/mm^2 and up to 9.8 $x10^3$ N/mm^2, fb above 8.5 N/mm^2 and up to 12 N/mm^2.

Four grade stresses have also been specified with strength ratios of 80, 63, 50 and 40%. For example, a species at moisture content >18% belonging to strength group N_1 of 80% grade has a bending strength of 0.8×28 $N/mm^2 = 22.4$ N/mm^2. The grade stresses for the seven strength groups have been tabulated in the Nigerian Code of Practice for Timber Design. The timber groupings based on strength are presented in Appendix A.1.

The grade standard established by the Kenya Bureau of Standards is different from that of Nigeria as shown in Table 5.4. The S-75 and S-50 grades are commonly specified for use in building construction. However, there are several tree species found across Africa, many of which are used only in very local areas. To

Table 5.4 Kenyan timber grades and application

Grade	Applications
F	Furniture, high-class joinery
GJ	General joinery
S-75	Structural grade, having a value of 75% of basic stress
S-50	Structural grade, having a value of 50% of basic stress
C	A general construction grade for non-stressed construction
L	A low grade for low quality work

Source Mrema et al. (2011)

Table 5.5 Guide to basic working-stress values and modulus of elasticity for timber

Strength group	Green density (kg/m^3)	Density at 12% moisture content (kg/m^3)	Bending and tension parallel to grain (N/mm^2)	Compression parallel to grain (N/mm^2)	Compression perpendicular to grain (N/mm^2)	Shear parallel to grain (N/mm^2)	Modulus of elasticity (N/mm^2)
1	<520	400	10	2.5	0.6	1.0	4000
2	521–650	401–500	15	10	1.2	1.3	6000
3	651–830	501–640	20	13	2.0	1.9	7500
4	831–1040	641–800	30	20	3.2	2.4	9000
5	>1041	>801	50	29	5.0	3.2	10,500

Source Mrema et al. (2011)

obtain approximate working-stress data for these indigenous species, their densities may be used to group them. If the density is not known, a simple experiment can be performed quite easily. A bucket, a graduated cylinder (millilitres) and an accurate scale for weighing a sample of the wood are the only materials required. The procedure is as follows:

(i) Weigh and measure the dimensions of the wood sample.
(ii) Place the bucket on a level surface and fill to the rim with water.
(iii) Carefully submerge the wood sample and then remove.
(iv) Refill the bucket from the graduated cylinder, noting the amount of water needed to refill the bucket.
(v) Compute the value of density = weight/volume (kg/m^3).
(vi) Place the species in an appropriate strength group using the density column for a green or dry sample (see Table 5.5, column 2 or 3).

The values shown in Table 5.5 are basic working stresses. For design purposes, these values may have to be adjusted for grade, moisture content, load duration, exposure and use of the structure.

5.4 Laminated Solid Wood Products

The array of laminated solid wood products available in the global market includes Glued-laminated timber, nail-laminated timber, cross-laminated timber, and dowel-laminated timber. Interestingly, there has been continued product innovation in the laminated wood product industry such that today, there are both old and new laminated products. While glued-laminated timber and nail-laminated timber belong to the second generation of wood products, both cross-laminated timber and dowel-laminated timber are newer, fourth-generation wood products. Glued-laminated timber (GLT), otherwise known as glulam (Fig. 5.12), is fabricated by gluing thin laminations of wood together in such a way to produce wood members of practically any size and length. Glulam members are small pieces of wood glued

Fig. 5.12 Glulam

together either in straight or in curved form, with the grain of all the laminations essentially parallel to the length of the member. Thus, laminated wood is basically different from plywood in which the grain direction of adjacent plies is at right angles. Though not yet in common use in many parts of tropical Africa due to several reasons, particularly relatively high cost, glulam is frequently used in other parts of the world in place of sawn timber for truss members, beams, stringers, columns and arches. It has the advantage of higher working stresses, of being available in larger sizes and longer lengths than sawn timber and of reducing to a minimum the problems of shrinkage and the resultant secondary stresses in joints. The lumber used in laminating is usually seasoned and accurately surfaced before being glued.

Nail-laminated timber (NLT) is produced when timber pieces are mechanically laminated to create a solid structural element. NLT is created by placing dimension lumber (nominal 50–100 mm thickness and 100–150 mm width) on edge and fastening the individual laminations together with nails. NLT is typically used as floors and roofs and can also be used for walls, elevator shafts and stair shafts. If plywood/OSB is added to one face, it can provide in-plane shear capacity, allowing the product to be used as a shear wall or diaphragm. NLT offers a consistent and attractive appearance for decorative and exposed-to-view applications. It requires no necessarily unique manufacturing facility and can be fabricated with local dimension lumber for use in applications across sectors and structure types (Fig. 5.13).

Cross-laminated timber (CLT) panels consist of several layers of structural lumber boards stacked crosswise (typically at 90°) and glued together on their wide faces and, sometimes, also on the narrow faces as shown in Fig. 5.14. A cross-section of a CLT element has at least three glued layers of boards placed in orthogonally alternating orientation to the neighbouring layers. In special

Fig. 5.13 Nail-laminated timber

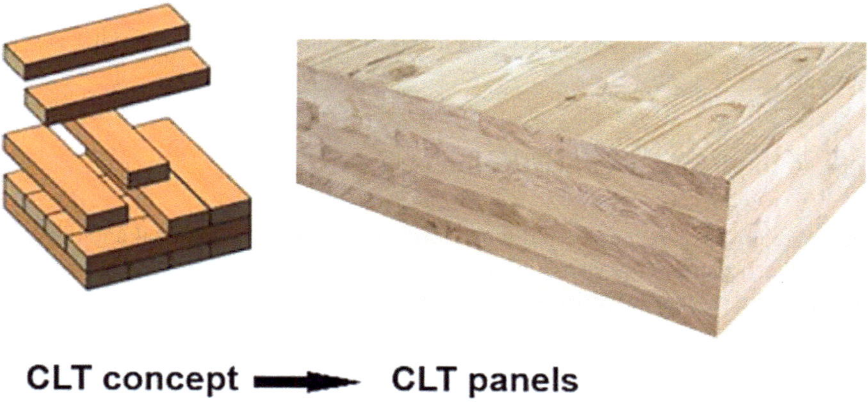

Fig. 5.14 Cross-laminated timber manufacture

configurations, consecutive layers may be placed in the same direction, giving a double layer (e.g. double longitudinal layers at the outer faces and additional double layers at the core of the panel) to obtain specific structural capacities. CLT products are usually fabricated with three to seven layers and even more in some cases. CLT panels are typically used as load-carrying plate elements in structural systems such as walls, floors and roofs. Lumber in the outer layers of the panels used as walls is normally oriented parallel to vertical loads to maximize the wall resistance. Likewise, for floor and roof systems, the outer layers run parallel to the major span direction.

Dowel-laminated timber (DLT) panels (Fig. 5.15) are made from lumber pieces stacked like NLT, but friction-fitted together with dowels. Hardwood dowels hold each board side-by-side, similar to how nails work in an NLT panel, and the friction fit lends some dimensional stability to the panel. DLT panels may be processed

Fig. 5.15 Dowel-laminated timber

using Computer Numerical Control (CNC) machinery, creating a high tolerance panel, which can also contain pre-integrated electrical conduits and other service runs.

5.5 Wood-Based Panel Products

These are products manufactured with wood as the basic raw material and designed with rather specific characteristics such as resistance to differential shrinkage and swelling, fire resistance, resistance to weathering. The most common panel products in tropical Africa are plywood, resin-bonded particleboards and fibreboards (shown in Fig. 5.16a–c). Plywood is produced by laying an odd number of wood veneers at right angles to each other and gluing them together under heat and pressure. It is a more dimensionally stable wood product than lumber and is typically produced as flat sheets. Its structural applications include floor underlayment, floor sheathing. Its non-structural applications include wall panelling (decorative) and form work (functional).

The term particle is a generic name for all lignocellulosic elements from which particleboard is made. The particles can be as coarse as pulp chips or as fine as sander dust. Some particles are by-products of wood processing operations, while others are principally generated through hammer milling for particleboard production. The reduction of feedstock into a given particle type depends on the form of raw material. Particleboard is a generic term for a panel manufactured from lignocellulosic material primarily in the form of discrete pieces or particles, as distinguished from fibres, combined with a synthetic resin or other suitable binder and bonded together under heat and pressure in a hot press. In general, lower density wood species (<500 kg/m^3) produce boards with superior strength properties. This is because denser species are often difficult to compress into well-bonded boards. However, the denser materials can be blended with lighter

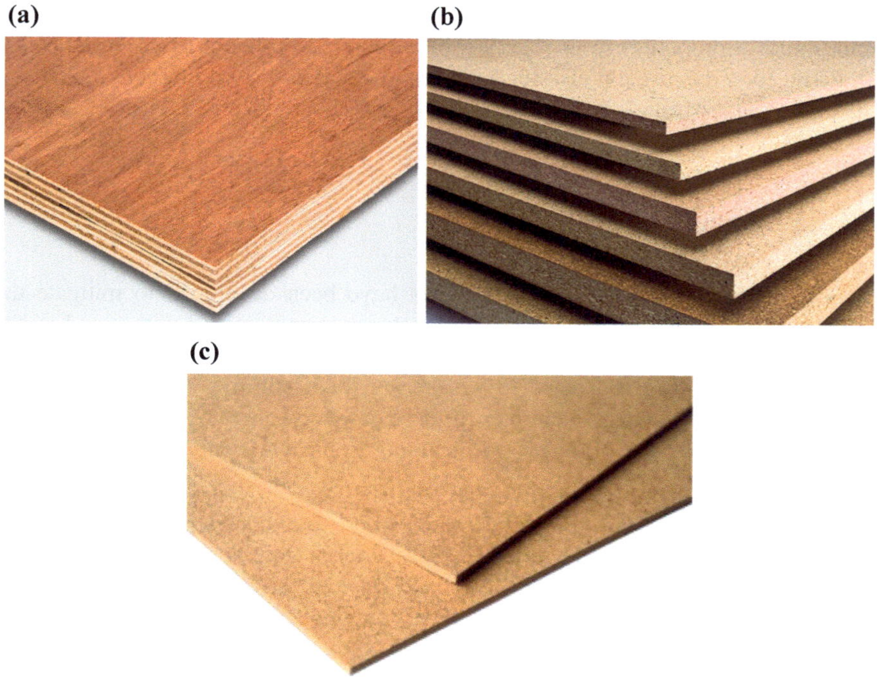

Fig. 5.16 **a** Plywood **b** Particleboards **c** Fibreboards

ones to produce acceptable boards. Particleboard products are further defined by the method of pressing. Flat-platen-pressed particleboard is produced when the pressure is applied in the direction perpendicular to the faces, while extruded particleboard is produced with pressure applied parallel to the faces.

Particle geometry (shape and size) is a prime consideration in particleboard manufacturing as it affects mechanical strength, i.e. bending, tension parallel and perpendicular to the board surface, screw and nail holding, properties. It also affects the water resistance ability, stability and surface smoothness of the board. The machining characteristics of the particleboard are also affected. The most commonly used particleboard is the flat-platen-pressed type. This product is utilized in diverse forms in building construction, furniture making, packaging, etc, due to its

Table 5.6 Density classification of particleboard

Class	Density (kg/cm^3)
Low density	<0.5
Medium density	0.59–0.80
High density	>0.80

desirable characteristics including availability in large sheets, smooth surfaces, uniformity in properties from sheet to sheet and freedom from localized defects.

Particleboard is usually classified on density basis as shown in Table 5.6. Board thickness ranges from 9.5 to 19 mm. The typical board size is 2.44 m × 2.44 m.

5.6 Engineered Wood Products

Over the years, new wood-based products have been developed to mitigate the inadequacies of natural wood. These products are manufactured with wood as the basic raw material, but are designed with rather specific characteristics such as dimensional stability, resistance to weathering, high stiffness and/or strength, or multidimensional stiffness. Examples of these products, many of which are still not available in many parts of the world including tropical Africa, are as follows:

(i) **Oriented Strand Board (OSB)**, which is a non-veneer structural panel, manufactured from thin reconstituted wood strands or wafers (Fig. 5.17).

(ii) **Laminated Veneer Lumber (LVL)**, which is similar in certain respects to glulam and plywood. It is fabricated by laminating thin sheets of veneer (0.25–0.4 cm thick) bonded together with durable adhesive. The veneer plies are usually laid up with all the wood fibres running in one direction and aligned parallel to the length of the member (Fig. 5.18).

(iii) **Parallel Strand Lumber (PSL)**, which is a lumber composite material produced from long strands of wood material. The strands are dried, coated with a waterproof adhesive and bonded together under pressure and heat. The strands are usually aligned so that the wood grain is parallel to the length of the member (hence the name) (Fig. 5.19).

(iv) **Laminated Strand Lumber (LSL), which** is a non-veneer panel product manufactured by compressing wood strands of between 0.76 and 1.3 mm thickness coated with adhesive under heat. It is different form PSL in terms

Fig. 5.17 Oriented strand board

Fig. 5.18 Laminated veneer lumber

Fig. 5.19 Parallel strand lumber

of strand size, type of adhesive used and method of consolidation. For example, the strands used in PSL fabrication are typically about 3.2 mm thick (Fig. 5.20).

LVL, PSL and LSL are collectively known as *Structural Composite Lumber* (SCL).They have gained substantial acceptance in the residential building construction industry in the last couple of decades. They are

Fig. 5.20 Laminated strand lumber

Fig. 5.21 Wood–cement composite panels

generally used for floor and roof sheathing and as walling materials. The development of these engineered wood products has made possible the use of lower quality and non-traditional forest resources to create high-performance structural materials.

(v) **Wood–cement Composite panels**, a set of lightweight concrete products (shown in Fig. 5.21) in which wood particles, shavings, strands or chips serve as aggregate in cement–water mixtures.

The processes involved in the manufacture of composite wood–cement panels include the following:

- **Wood Material Preparation**: The wood could be obtained as round logs, from thinnings, logging and wood processing residues such as sawdust, shavings and chips. Solid wood requires debarking and ripping into slabs, a process that would be unnecessary for sawdust, wood flakes or chips.
- **Size Reduction**: At this stage, the wood slabs are reduced into either flakes (30–50 mm long and 0.2–0.6 mm thick) or finer sawdust particles. The final particle size to which the material is reduced depends the ultimate end-use requirement of the board. Longer and thinner flakes are usually stronger, stiffer and more dimensionally stable.

- **Mixing/Blending**: The flakes or particles are thoroughly mixed with water containing additive in a predetermined mixing ratio. Cement is then slowly added while mixing continues until a homogeneous wood/cement mix of the desired consistency is formed.
- **Forming**: This is the process of laying down a mat of the blended particles. The mats are usually formed onto metal 'caul plates'.
- **Pressing**: Pressing involves consolidating the mat. The acceptable mats are stacked up into the required clamping sizes and passed on to the press where it is tamped to the predetermined thickness.
- **Maturation**: The panels are removed from the press and stored for at least 28 days to allow for further curing of the cement. By the 28th day, the panel would have attained maximum strength.
- **Trimming**: The panels are trimmed to the required sizes and finally stored or shipped as the case may be.

The admirable properties of wood–cement panel products that recommend them for use as building material include relatively high strength-to-weight ratio; durability; stability, i.e. high resistance to moisture uptake; nailability; ease of sawing; excellent insulation against noise and heat; and high resistance against fire, insect and fungus attack. The panels are also environmentally friendly since they do not emit gasses or leak harmful chemicals. Besides, they are coatable with paints and plasters and lend themselves to modular construction. These products have another special appeal by virtue of the fact that they can be, and mostly are made, from wood and non-commercial or low-value tree species. Because they are produced in panel form, they can be substituted for or used in combination with other materials commonly used in building construction. Over the years, many tropical hardwood species have been successfully tested for wood–cement panel production. These include Afara (*Terminalia superba*), *Triplochiton scleroxylon*, *Mitragyna ciliata*, *Ceiba pentandra*, *Tectona grandis*, *Melia composita*, *Gmelina arborea*, *Antiaris africana* and *Brachystegia kennedyi*.

A global market already exists for wood–cement panel products. They are employed as construction materials for interior/exterior wall cladding, roofing, ceiling and shuttering for bungalows and high-rise buildings. They have become widely acceptable in many countries in Europe, Asia, North and Central America, and the Middle East, including Germany, England, Japan, Indonesia, United State of America, Brazil, Iran and Saudi Arabia. They have long been in use in these parts of the world, both to satisfy emergency situations (e.g. to provide housing for people affected by natural disasters), and for regular building construction. Because they are lighter in weight than most other cementitious materials and are produced in panel form, the material can be easily packaged and shipped to different parts of the world. A primary concern in introducing these products into international markets, however, is compliance with local building codes. Most local codes in European, North and Central American countries specify design loads, wind resistance, fire rating and seismic strength. However, there are no specific limitations on the use of cement-bonded wood composites.

Fig. 5.22 Wood–plastic composites

(vi) **Wood–Plastic Composites (WPCs)**, shown in Fig. 5.22, which are produced by thoroughly mixing ground wood particles and heated thermoplastic resin. The most common method of production is to extrude the material into the desired shape, though injection moulding is also used. WPCs may be produced from either virgin or recycled thermoplastics. These composites are more environmentally friendly and require less maintenance than the alternatives of solid wood treated with preservatives or solid wood of rot-resistant species. One advantage of WPC over wood is the ability of the material to be moulded to meet almost any desired shape. A WPC member can be bent and fixed to form strong arching curves.

Another major selling point of these materials is their lack of need for paint. They are manufactured in a variety of colours, but are widely available in greys and earth tones. The most widespread use of WPCs in North America is in outdoor deck floors, but they are also used for railings, fences, landscaping timbers, cladding and siding, park benches, moulding and trim, window and door frames and indoor furniture. However, WPCs have a lower strength and stiffness than wood, and they are also visco-elastic, i.e. experience time and temperature-dependent behaviour. The wood particles are susceptible to fungal attack, though not as much so as solid wood, and the polymer component is vulnerable to ultraviolet degradation. Some WPC formulations are also sensitive to staining from a variety of agents. The polymers and adhesives added make wood–plastic composite difficult to recycle again after use. They can however be recycled easily in a new wood–plastic composite.

5.7 A Glimpse into the Future

The most potentially revolutionary development in wood utilization will be the creation, through nanotechnology, of new materials that rival the strength and lightness of metals and plastics. Nanotechnology is envisaged as the main driving technology of the future in the wood products industry. It promises revolution in two areas: things done *to* wood products (preservatives, sealants, adhesives, etc.) and things done *with* wood fibres (new materials to replace non-renewable metals, ceramics and plastics). With further breakthroughs in technology on the nanoscale, new wood products not yet on the horizon will be developed. Potential uses for nanotechnology include developing intelligent wood and paper-based products with an array of nanosensors built into measure forces, loads, moisture levels, temperature, pressure, chemical emissions, attack by wood decaying fungi, etc. Nanotechnology will result in a unique next generation of wood-based products that have hyperperformance and superior serviceability in severe environments, strength properties now only seen with carbon-based composite materials, durability in service and biodegradability useable service life.

Practice Questions

1. What are the advantages of wood as a structural material?
2. List and describe five forms in which wood and wood products are used for structural purposes.
3. Describe five types of wood products used in building construction.
4. If you were asked to choose between lumber and logs as a construction material, give five reasons why you would select logs instead of sawn wood.

References

Adefisan OO, Ogundipe A (2016) Recent developments in the manufacture of wood-based composites. In: Olorunnisola AO (ed) Renewable natural resources engineering: essays in honour of Canon Prof. E. Babajide Lucas. Zenith Book House Publishers, Ibadan, pp 151–164

Alberto MM, Mougel Zoulalian A (2000) Compatibility of some tropical hardwood species with Portland cement using isothermal calorimetry. Forest Prod J 50(9):83–88

Badejo SOO (1986) Dimensional stability of cement-bonded particleboard from eight tropical hardwoods grown in Nigeria. Niger J Forest 16(1/2):11–19

Badejo SOO (1987) An investigation on the influence of cement binder content on properties of cement-bonded particleboard from four tropical hardwoods. Malays Forester 50(1):107–120

Badejo SOO (1988) Influences of flake geometry on properties of cement-bonded particleboard from mixed tropical hardwoods. Wood Sci Technol 22:357–370

Badejo SOO (1989) Influences of pre-treatment temperature and additive concentration on the properties of cement-bonded particleboard from plantation grown tropical hardwood. Trop Sci 29:285–296

Bodig J, Jayne BA (1982) Mechanics of wood and wood composites. Van Nostrand Reinhold Company, New York, USA

Faherty KF, Williamson TG (1999) Wood engineering and construction handbook, 3rd edn. McGraw-Hill Handbooks, New York, USA

Forest Products Laboratory (2010) Wood handbook—wood as an engineering material, reneral technical report FPL-GTR-190. The U.S. Department of Agriculture, Forest Service, Forest Products Laboratory, Madison, WI

Kollman FFP, Cote WA Jr (1968) Principle of wood science and technology. Springer, Berlin

Lucas EB, Olorunnisola AO (2002) Wood processing and utilization in Nigeria: the present situation and future prospects. In: Ajav EA, Raji AO, Ewemoje TA (eds) Agricultural engineering in Nigeria: 30 years of University of Ibadan experience. The Department of Agricultural Engineering. University of Ibadan, Nigeria, pp 98–109

Mijinyawa Y, Alabadan BA (2016) Structural utilisation of wood. In: Olorunnisola AO (ed) Renewable natural resources engineering: essays in honour of Canon Prof. E. Babajide Lucas. Zenith Book House Publishers, Ibadan, pp 139–150

Mrema GC, Gumbe LO, Chepete HJ, Agullo JO (2011) Rural structures in the tropics: design and development. FAO, Rome

NCP (2005) Nigerian code of practice for structural timber design. The Standards Organisation of Nigeria

Nelson S (1997) Structural composite lumber. In: Smulski S (ed) Engineered wood products, a guide for specifiers, designers and users. PFS Research Foundation, Madison, Wisconsin

Olorunnisola AO (2009) Suitability of four Nigerian hardwoods for wood-cement particleboard production. J Forest Res Manage 6:10–17

Olorunnisola AO, Omoniyi TE (2016) Wood conversion and processing in Nigeria: a status report. In: Olorunnisola AO (ed) Renewable natural resources engineering: essays in honour of Canon Prof. E. Babajide Lucas. Zenith Book House Publishers, Ibadan, pp 101–121

Olufemi B (2016) Wood utilisation pattern in Nigeria—an overview. In: Olorunnisola AO (ed) Renewable natural resources engineering: essays in honour of Canon Prof. E. Babajide Lucas. Zenith Book House Publishers, Ibadan, pp 82–90

Smulski S (1997) Engineered wood products- a guide for specifiers, designers and users. PFS Research Foundation, Madison, Wisconsin

Walker JCF, Butterfield BG, Langrish TAG, Harris JM, Uprichard JM (1993) Primary wood processing principles and practice. Chapman and Hall Publishers, London

Zylkowski S (2002) Introduction to wood as an engineering material. In: Williamson TG (ed) APA engineered wood handbook: 1.1-1.25, McGraw-Hill Publishers, New York

Chapter 6
A Review of the Basic Theory of Structures

6.1 Forces in Structures

A structure can be defined as a body that is capable of resisting applied loads without appreciable deformation of one part relative to another. The function of a structure is to transmit forces from one point in space to another. In general, the force actions on any structural element may include forces normal to element axis, i.e. shear forces; force parallel to element axis, i.e. axial or normal force, bending actions about three axes, i.e. bending moments and twisting moment. A characteristic set of internal stresses is associated with each of these basic forces. One of the fundamentals in the study of structures is to find these characteristic stress patterns and deformations and to relate them to the basic force actions given. The main objective of structural analysis and design, therefore, is to develop a structure that is satisfactory in service in terms of strength, i.e. it must not collapse when loads are applied, and stiffness, i.e. the deformations must not be excessive. The design of any structure requires many detailed computations. Some of these are of a routine nature. An example is the computation of section properties for standard sized species and grades of dimension timber. The wider availability of the computer in the last two decades has resulted in its rapid adoption for structural design in replacement for manual computations and has helped in reducing the amount of time required for both analysis and design.

6.2 Basic Elements of a Framed Structure

There are two broad sub-divisions of structures, i.e. mass structures which resist applied loads by virtue of their weight, e.g. dams, and framed structures which resist applied loads by virtue of their geometry, e.g. furniture items, framed buildings, aeroplanes, and ships. Wood is seldom used in the design of mass

A.O. Olorunnisola, *Design of Structural Elements with Tropical Hardwoods*,
DOI 10.1007/978-3-319-65343-3_6

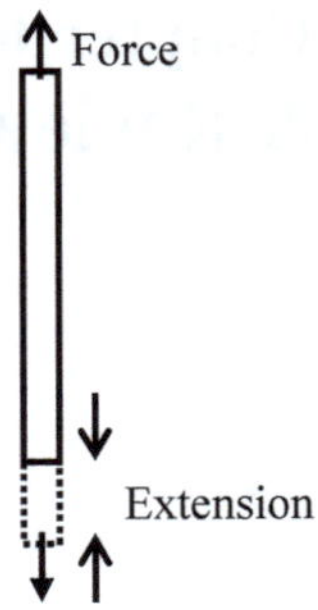

Fig. 6.1 A rod under loading

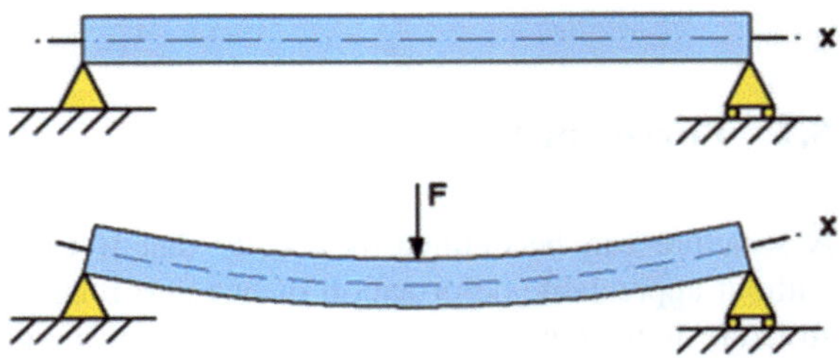

Fig. 6.2 A beam under loading

structures. We are, therefore, concerned with the design of wooden members for framed structures. There are three basic elements of a framed structure. They are:

(i) **The rod**: a member subjected to loading along the axis (tensile for a tie, compressive for a column or strut) and the deformation under load is a simple change in length as shown in Fig. 6.1.
(ii) **The beam**: A member supported at one or more points in its length and carrying load(s) normal to the member axis (Fig. 6.2). Its characteristic deformation by bending action is a plane curvature.
(iii) **The Slab or Plate**: A slab sustains a load normal to its axis and deforms like a beam, but into a dished shape having two-way curvature.

6.3 Types of Joints in a Framed Structure

A framed structure comprises members joined together. The two basic types of joints used are ***stiff*** and ***pinned*** joints. In a stiff joint (Fig. 6.3a, b), a flexure of one member meeting at the joint has an effect on the other members, e.g. a plywood gusset plate joint (Fig. 6.3c) or wood–concrete joint. If a joint is perfectly stiff, then the angle between the members is unaltered while the joint rotates. In a pinned joint (Fig. 6.4a, b), the flexure of a member on one side of the joint does not affect other members meeting at that joint, e.g. a nailed joint (Fig. 6.4c). Hence, while a stiff joint can transmit bending from one side to another, a pinned joint cannot.

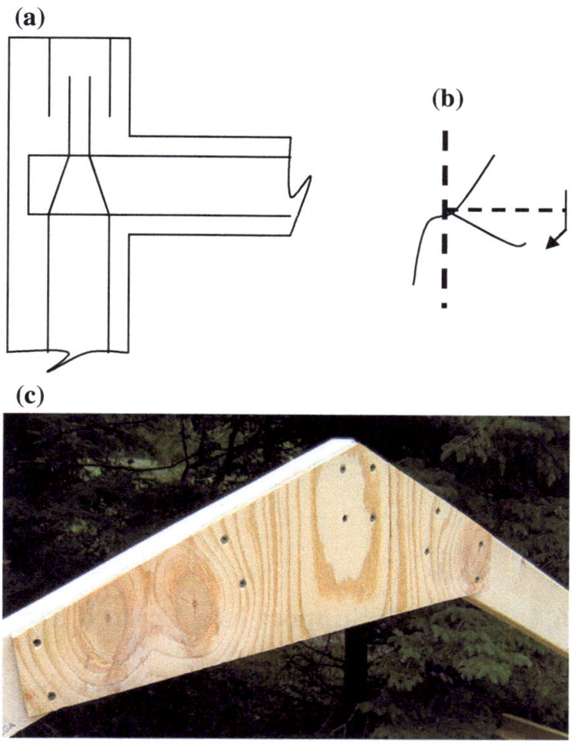

Fig. 6.3 **a** A rigid joint. **b** A stiff joint in rotation. **c** A rigid plywood gusset plate joint

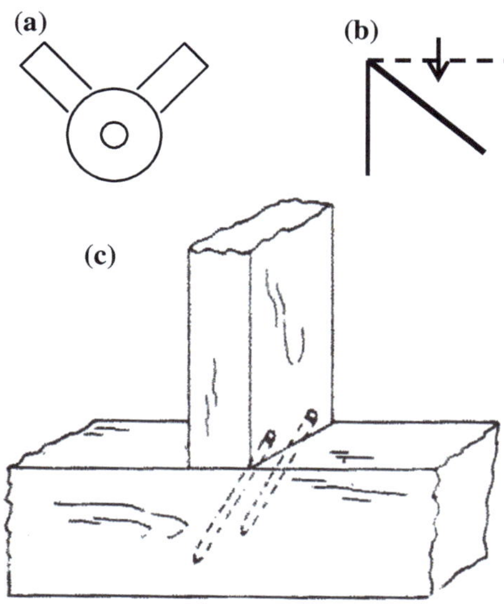

Fig. 6.4 **a** A pinned joint. **b** Flexure of a pinned joint. **c** A pinned nailed joint

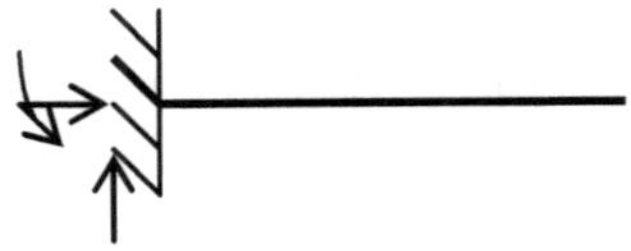

Fig. 6.5 A fixed end or built-in support

Fig. 6.6 A pin support

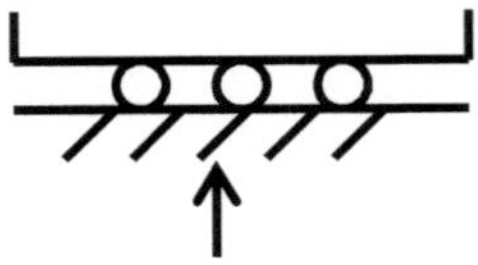

Fig. 6.7 A roller support

6.4 Types of Support in a Framed Structure

The load applied to a framed structure is transmitted to supports which will supply the necessary reactive forces to maintain equilibrium. In a plane structure (i.e. a structure having all its members and loads in one plane), the different types of support are shown in Figs. 6.5, 6.6, and 6.7.

A fixed end or built-in support, shown in Fig. 6.5, is capable of supplying three reactive forces: horizontal, vertical and fixing moment. The pin support, shown in Fig. 6.6, can supply two reactive forces: horizontal and vertical, while the roller bearing, shown in Fig. 6.7, can only supply a vertical reaction.

6.5 Equations of Equilibrium

Since a structure is a body in static equilibrium without any dynamic effects, any force action in space can be replaced by a force at the origin of the three reference axes (x, y, z), together with a moment about one of these axes. Then, for equilibrium, the resultant action in any direction must be zero giving the six equations of equilibrium:

$$\begin{aligned} \sum P_x = 0 \quad \sum M_x = 0 \\ \sum P_y = 0 \quad \sum M_y = 0 \\ \sum P_Z = 0 \quad \sum M_Z = 0 \end{aligned} \tag{6.1}$$

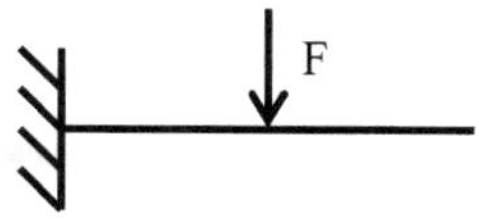

Fig. 6.8 A cantilever

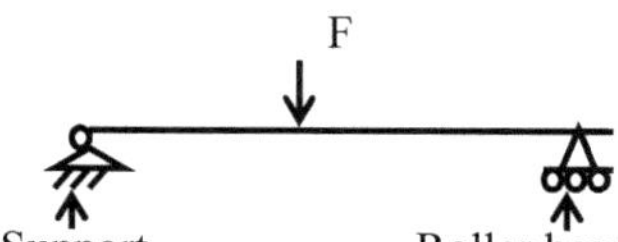

Fig. 6.9 A simply supported beam

In the case of a plane frame lying in the *xy*—plane, these equations reduce to:

$$\sum P_x = 0; \sum P_y = 0; \sum M_Z = 0 \tag{6.2}$$

Some structures can be completely analysed by the use of these equations. Such structures are known as ***statically determinate*** or simply ***determinate structures***. Structures which cannot be analysed solely by the application of these equations are known as ***statically indeterminate***, ***hyperstatic*** or ***redundant structures***. In a simple beam, the condition for determinacy is that the supports must be such that there are not more than three reactive forces. In other words, they must be either built-in at one end with no support whatsoever at the other (cantilever), or the supports must consist of a pin at one end and a roller bearing at the other (see Figs. 6.8 and 6.9).

6.6 Basic Properties of Sections

Since framed structures resist loads by virtue of their geometry, it is important to understand the basic properties of common geometrical sections employed in wood structural design. A brief review follows.

6.6.1 Moments of Area

The first moment of an element of area about any axis in the plane of the area is given by the product of the area of the element and the perpendicular distance between the element and the axis; that is, the first moment dQ_x of the element da about the *x*-axis is given by $dQ_x = yda$. About the *y*-axis the first moment is $dQy = xda$. For a finite area, ($Qx = \int dQx$), i.e. the first moments of area for a finite area about any axis in the plane of the area is given by the summation of the first moments about that same axis of all the elements of area contained in the finite area.

6.6.2 Centroid of an Area

The centroid of an area can be computed using the following equations:

$$\overline{X} = \frac{\int xda}{A} = \frac{Qy}{A}; \quad \bar{y} = \frac{\int yda}{A} = \frac{Qx}{A}$$

$$\overline{X} =, \frac{\sum xiAi}{\sum Ai} \quad \bar{y} = \frac{\sum yiAi}{\sum Ai}$$

where **A** denotes area. The centroid of an area is the point at which the area might be considered to be concentrated and still leave unchanged the first moment of the area about any axis. In a symmetrical figure such as a circle or square, the centroid coincides with the geometric centre of the figure. Note that the first moment of area (statistical moment of the area) will be zero whenever the reference axis passes through the centroid.

6.6.3 Moments of Inertia of Various Sections

The moment of inertia of a body is the product of the area and the square of its distance from the axis of reference. The following formulas are used for computing the moment of inertia, ‘I’, for different cross-sections:

(a) A rectangle of depth ‘d’ and width ‘b’ about a centroidal axis parallel to the base:

$$I = \frac{bd^3}{12}$$

(b) A rectangle of depth ‘d’ and width ‘b’ about the base of the rectangle:

$$I = \frac{bd^3}{3}$$

(c) A rectangle of depth ‘d’ and width ‘b’ from which a coaxial symmetrical rectangular hole has been punched out:

$$I = \frac{BD^3}{12} - \frac{bd^3}{12}$$

(d) A circular cross-section of diameter 'd' about a centroidal axis lying in its plane:

$$I = \frac{\Pi D^4}{64}$$

(e) An I cross-section:

$$I = \frac{BD^3}{12} - \frac{bd^3}{12}$$

Note: Units of moment of inertia are the 4th power of a length, mm^4, m^4.

The moment of inertia of a composite area is the summation of the moments of inertia of the component areas making up the whole. This frequently eliminates the necessity for integration if the area can be broken down into rectangles, triangles, circles, for each of which the moment of inertia is known.

6.6.4 Section Modulus

The section modulus of a body is the moment of inertia divided by the distance from the neutral axis. The following formulas are used for computing the section modulus, 'S', for different geometrical cross-sections, i.e.

(a) For a rectangular cross-section of depth 'd' and width 'b':

$$S = bd^2/6$$

(b) For a circular cross-section of diameter 'd':

$$S = \pi d^3/32$$

(c) For an I cross-section of flange width 'B' and thickness 'D' and a web width 'b; and thickness d:

$$S = \left[BD^3/12 - bd^3/12\right]/D/2$$

The units of section modulus are mm^3, cm^3 and m^3.

6.6.5 *Radius of Gyration*

If the moment of inertia of an area A about the x-axis is denoted by I_x, and the area by A, then the radius of gyration 'r_x' is defined as the square root of the moment of inertia divided by the cross-sectional area, i.e.

$$r_x = \sqrt{I_x}/A$$

Similarly, the radius of gyration with respect to the y-axis 'r_y' is given by:

$$r_y = \sqrt{I_y}/A$$

where A = cross-sectional area and I_y = moment of inertia about the y-axis.

Recall also that the units of radius of gyration are mm, cm and m.

The cross-sectional area, section moduli, moments of inertia and radii of gyration of lumber of different nominal sizes available in the Nigerian and other timber markets in West Africa loaded about the strong axis (x-axis) and the weak axis (y-axis) are shown in Tables 6.1 and 6.2.

Worked Examples

1. A rectangular wooden beam 8 cm wide and 16 cm deep determines the following:

   (a) Moments of inertia about centroidal axes parallel to the two sides
   (b) The radius of gyration about the same axes, and
   (c) The section moduli.

Solution

(a) Moments of Inertia

$$I_{xx} = \frac{bh^3}{12} = \frac{(8)(16)^3}{12}$$
$$= 2730.68\,\text{cm}^4$$
$$I_{yy} = \frac{hb^3}{12} = \underline{(16)(8)^3} = 682.67\,\text{cm}^4$$

(b) Radii of gyration are:

$$r_{xx} = \sqrt{I_{xx/A}} = \sqrt{2730.67}/8 \times 16 = 4.62\text{ cm}$$
$$r_{yy} = \sqrt{I_{yy/A}} = \underline{\sqrt{682.67}}/8 \times 16 = 2.31\text{ cm}$$

Table 6.1 Section properties of construction lumber loaded along the strong axis

Thickness (*B*) (mm)	Width *D* (mm)	Area (m^2)	*S* (mm^3)	*I* (mm^4)	*r* (mm)
50	75	3750	46,875.0	1,757,812.5	21.7
50	100	5000	83,333.3	4,166,666.7	28.9
50	125	6250	130,208.3	8,138,020.8	36.1
50	150	7500	187,500.0	14,062,500.0	43.3
50	175	8750	255,208.3	22,330,729.2	50.5
50	200	10,000	333,333.3	33,333,333.3	57.7
50	225	11,250	421,875.0	47,460,937.5	65.0
50	250	12,500	520,833.3	65,104,166.7	72.2
50	300	15,000	750,000.0	112,500,000.0	86.6
75	100	7500	125,000.0	6,250,000.0	28.9
75	125	9375	195,312.5	12,207,031.3	36.1
75	150	11,250	281,250.0	21,093,750.0	43.3
75	175	13,125	382,812.5	33,496,093.8	50.5
75	200	15,000	500,000.0	50,000,000.0	57.7
75	225	16,875	632,812.5	71,191,406.3	65.0
75	250	18,750	781,250.0	97,656,250.0	72.2
75	300	22,500	1,125,000.0	168,750,000.0	86.6
100	100	10,000	166,666.7	8,333,333.3	28.9
100	150	15,000	375,000.0	28,125,000.0	43.3
100	200	20,000	666,666.7	66,666,666.7	57.7
100	250	25,000	1,041,666.7	130,208,333.3	72.2
100	300	30,000	1,500,000.0	225,000,000.0	86.6
150	150	22,500	562,500.0	42,187,500.0	43.3
150	200	30,000	1,000,000.0	100,000,000.0	57.7
150	250	37,500	1,562,500.0	195,312,500.0	72.2
150	300	45,000	2,250,000.0	337,500,000.0	86.6
200	200	40,000	1,333,333.3	133,333,333.3	57.7
250	250	62,500	2,604,166.7	325,520,833.3	72.2
300	300	90,000	4,500,000.0	675,000,000.0	86.6

Note S—section modulus, *I*—moment of inertia, *r*—radius of gyration

(c) Section moduli are:

$$Z_{xx} = \frac{I_{xx}}{y} = \frac{2730.67}{(16/2)} = 341.33\ \text{cm}^3$$

$$Z_{yy} = \frac{I_{yy}}{12} = \frac{682.67}{(8/2)} = 170.67\ \text{cm}^3$$

Table 6.2 Section properties of construction lumber loaded along the weak axis

Thickness (*B*) (mm)	Width *D* (mm)	Area (m^2)	*S* (mm^3)	*I* (mm^4)	*r* (mm)
50	75	3750	31,250.0	781,250.0	14.4
50	100	5000	41,666.7	1,041,666.7	14.4
50	125	6250	52,083.3	1,302,083.3	14.4
50	150	7500	62,500.0	1,562,500.0	14.4
50	175	8750	72,916.7	1,822,916.7	14.4
50	200	10,000	83,333.3	2,083,333.3	14.4
50	225	11,250	93,750.0	2,343,750.0	14.4
50	250	12,500	104,166.7	2,604,166.7	14.4
50	300	15,000	125,000.0	3,125,000.0	14.4
75	100	7500	93,750.0	3,515,625.0	21.7
75	125	9375	117,187.5	4,394,531.3	21.7
75	150	11,250	140,625.0	5,273,437.5	21.7
75	175	13,125	164,062.5	6,152,343.8	21.7
75	200	15,000	187,500.0	7,031,250.0	21.7
75	225	16,875	210,937.5	7,910,156.3	21.7
75	250	18,750	234,375.0	8,789,062.5	21.7
75	300	22,500	281,250.0	10,546,875.0	21.7
100	100	10,000	166,666.7	8,333,333.3	28.9
100	150	15,000	250,000.0	12,500,000.0	28.9
100	200	20,000	333,333.3	16,666,666.7	28.9
100	250	25,000	416,666.7	20,833,333.3	28.9
100	300	30,000	500,000.0	25,000,000.0	28.9
150	150	22,500	562,500.0	42,187,500.0	43.3
150	200	30,000	750,000.0	56,250,000.0	43.3
150	250	37,500	937,500.0	70,312,500.0	43.3
150	300	45,000	1,125,000.0	84,375,000.0	43.3
200	200	40,000	1,333,333.3	133,333,333.3	57.7
250	250	62,500	2,604,166.7	325,520,833.3	72.2
300	300	90,000	4,500,000.0	675,000,000.0	86.6

Note *S*—section modulus, *I*—moment of inertia, *r*—radius of gyration

2. Create a Microsoft Excel template to compute the area, section modulus and moment of inertia of a 75 × 100 mm lumber piece in edgewise and flatwise uses.

Solution Steps

1. Open an Excel worksheet.
2. Input the width and depth values to be used for computation in edgewise and flatwise configurations into the Excel cell as shown in Fig. 6.10a.

(a)

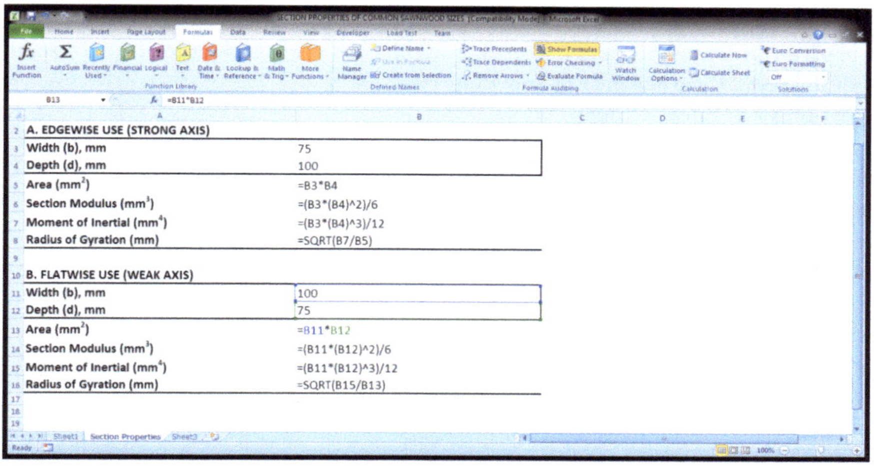

(b)

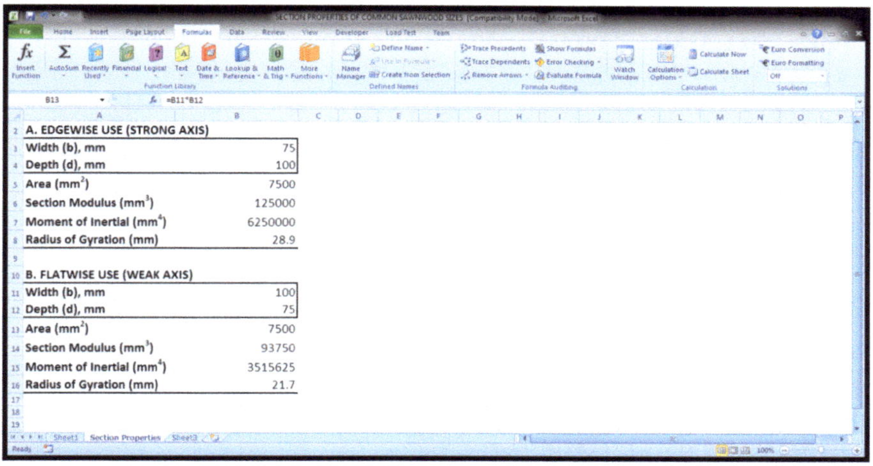

Fig. 6.10 a Excel worksheet showing the computations. **b** Excel worksheet showing the computation results

3. Input the formulas for computing area, section modulus moment of inertia and radius of gyration into the appropriate cells using the appropriate Excel cell references as shown in Fig. 6.10 and press 'Enter' each time.

The result of the computation is shown in the Excel worksheet presented in Fig. 6.10b.

3. Write a simple FORTRAN program to accept input from the user and compute the area, section modulus and moment of inertia of a rectangular lumber piece.

Solution

```
! Program on arrays to read lumber dimensions and compute sectional      areas,
section moduli and moments of inertia
!x = array of data values
! area = cross-sectional area of lumber sizes
!secmod = section modulus oflumber sizes in edgewise use
!mmt = moment of inertia of lumber sizes in edgewise use
real::x(100),y(100),area, secmod, mmt
integer::n,i
write(*,*) 'enter the number of lumber sizes to be handled'
read(*,*)n
write(*,*) 'enter lumber sizes, smaller dimensions, followed by larger di-
mensions'
    do i = 1,n
 read(*,*)x(i),y(i)
  end do
  write(*,5)'no','b','d'
 5 Format(A2,18X,A5,12X,A5)
      do i = 1,n
       write(*,10) i,x(i),y(i)
       10 format(I3,18X,F6.1,12X,F6.1)
         end do
    write(*,*)
    write(*,*)
    write(*,20)'n0','area','secmod','mmt of inertia'
 20 Format(A2,18X,A5,20X,A6,14X,A14)
    do i = 1,n
 area = x(i)*y(i)
 secmod = x(i)*y(i)**2/6
 mmt = x(i)*y(i)**3/12
      write(*,30)i,area,secmod,mmt
 30 Format(I3,4X,F20.2,4X,F20.2,4X,F20.2)
     end do
    end
```

Practice Questions

1. Distinguish between framed and mass structures.
2. Describe the three basic elements of a framed structure with the aid of suitable diagrams.
3. Sketch and describe three types of support commonly employed in a framed structure.

4. A number of 25 × 300 mm lumber pieces are to be used as decking materials in the design of a wood-framed building. If each piece of lumber is to be dressed down to 18.5 × 295.5 mm, compute the section modulus, moment of inertia and radius of gyration for each piece of lumber based on nominal and standard dressed sizes, respectively.
5. Determine the moment of inertia, section modulus and the radius of gyration of a 50-cm-diameter log.
6. For a rectangular wooden beam 12 cm wide and 36 cm deep, compute the following:
   (a) Moments of inertia about the centroidal axes parallel to the two sides.
   (b) The radii of gyration about the two axes.
   (c) The section moduli about the two axes.
7. Create a Microsoft Excel template for computing, tabulating and printing the area, section moduli and moments of inertia of structural members of (a) square and (b) circular cross-sections. Select 10 dimensions to test-run your template.

References

Khurmi RS, Khurmi N (2013) Strength of materials (mechanics of solids in S.I. units, international edition. In: S. Chand & Company PVT Ltd., (ed) New Delhi
Marshall WT, Nelson HM (1978) Structures. Pitman Publishing Ltd., London
NCP (2005) Nigerian code of practice for structural timber design. The Standards Organisation of Nigeria
Seward D (2003) Understanding structures, analysis, materials, design, 3rd edn. Palgrave Macmillan, New York
Singh S (1982) Strength of materials, 2nd revised edn. Vikas Publishing House PVT Ltd., U.P.
Shaeffer RE (1999) Elementary structures for architects and builders, 4th edn. Prentice-Hall Publishing Co., New Jersey

Chapter 7
Structural Load Computations

Structural design is the methodical investigation of the stability, strength and rigidity of structures. The basic objective is to produce a structure capable of resisting all applied loads without failure during its intended life. If the actual applied loads exceed the design specifications, the structure may fail to perform its intended function, with possible serious consequences. Therefore, before a structure can be designed with adequate strength characteristics, the loads to which it will be subjected must be determined. To do this, the designer should understand the nature and significance of the various types of loads that act on structures and then relate this information to all decisions on design, materials and construction methods. Not all loads can be predicted with complete accuracy, but with proper use of the information available, sufficiently accurate estimates can be made to insure adequate design. Many countries have structural design codes, codes of practice or other such technical documents which perform a similar function. It is necessary for a designer to become familiar with local requirements or recommendations in regard to correct practice. In this chapter, some examples are given, occasionally in a simplified form, in order to demonstrate procedures. They should not be assumed to apply to all areas or situations.

7.1 Types of Loads

The loads to which structures are subjected may be divided into two: vertical (gravity) loads and lateral forces. Vertical loads fall under two categories, i.e. dead loads and live loads. While dead loads are applied permanently, live loads tend to fluctuate with time. When these two are combined, they are referred to as the ***total load***.

A.O. Olorunnisola, *Design of Structural Elements with Tropical Hardwoods*,
DOI 10.1007/978-3-319-65343-3_7

7.1.1 Dead Loads

These are the actual weights of all the permanent components of a structure such as wood framing, roofing, plywood sheathing and insulation. On occasion, permanent equipment such as large air conditioners can be considered dead loads. Hence, dead loads are integral parts of the structure and the fixed equipment that are used in the construction of the structure. It is generally assumed that the designer can estimate dead load prior to undertaking the design and will recheck the dead load when the design is executed, making appropriate adjustments at that time. Weights of construction materials are usually specified in building codes. The dead loads of selected building materials are shown in Tables 7.1, 7.2 and 7.3.

In a typical wood structure, the dead load of the framing members usually represents a fairly small portion of the total design load. The dead load of a typical wood floor or roof system usually ranges between 335 and 958 Pa depending on the materials of construction, span lengths, and whether a ceiling is suspended below the floor or roof. For wood wall systems, dead load values might range between 192 and 958 Pa, depending on stud size and spacing and the type of wall sheathings used. Since most building dead loads are estimated as uniformly distributed loads in kg/m^2 or N/m^2, it is often convenient to convert the weights of framing members to these units. For example, if the weight per unit length of a wood framing member is known, and if the centre-to-centre spacing of parallel members is also known, the dead load in kg/m^2 or N/m^2 can easily be determined by dividing the weight per unit length by the centre-to-centre spacing. For example, if a set of 50×300 mm beams each of which weighs 64.5 N/m are spaced at 0.4 m on centre, the equivalent uniform load is $64.5\ \text{N/m} \div 0.4\ \text{m} = 161.25$ Pa.

Table 7.1 Unit weight per area of selected sheet materials

Material	Weight (kN/m^2)
Acoustic ceiling tiles	0.1
Asphalt (19 mm)	0.45
Aluminium roofing sheet	0.04
Glass (single glazing)	0.1
Plaster (per face of wall)	0.3
Plasterboard and skim	0.15
Rafters, battens and felt	0.14
Sand/cement screed (25 mm)	0.6
Slates	0.6
Steel roof sheeting	0.15
Timber floorboards	0.15
Vinyl tiles	0.05

Source Seward (2003)

Table 7.2 Weight per area of selected sheet materials

Material	Weight (N/m^2 or Pa)
Ceiling	
Acoustic fibre tile	47.9
Channel suspended system	47.9
Floors	
Cement finish, per inch of thickness	574.9
Ceramic or quarry tile, 18.75 mm	479
Lightweight concrete, per mm of thickness	287.4–479
Reinforced concrete, per mm of thickness	598.8
Stone, per mm of thickness	598.8
Linoleum, 6.25 mm	47.9
Terrazzo finish, 37.5 mm	910.1
Vinyl tile, 3.125 mm	67.1
Roofs	
Lumber sheathing, 25 mm nominal	119.8
Plywood sheathing, per mm of thickness	143.7
Gauge 14 corrugated aluminium	52.7
Gauge 16 corrugated aluminium	43.1
Gauge 18 corrugated aluminium	33.5
Gauge 20 corrugated aluminium	28.7
Gauge 14 corrugated galvanized steel	172.4
Gauge 16 corrugated galvanized steel	138.9
Gauge 18 corrugated galvanized steel	115
Gauge 20 corrugated galvanized steel	86.2
Corrugated asbestos 18.75 mm	143.7

Table 7.3 Unit weight per volume of selected building materials

Material	Weight (kN/m^3)
Aluminium	24
Bricks	22*
Concrete	24
Concrete blocks (lightweight)	12*
Concrete blocks (dense)	22*
Glass-fibre composite	18
Steel	18
Timber	6*

*Subject to considerable variation
Source Seward (2003)

Worked Examples on Dead Load Assembly

(i) Compute the total roof dead load for the following situation

Roofing	=310 Pa
Reroofing	=120
12.5 mm plywood	=72
Framing (estimate)	=153
Suspended ceiling (acoustic tile)	=96
Total Roof Dead Load	**=751 Pa, say 760 Pa**

(ii) Compute the total floor dead load for the following situation:

Floor covering (lightweight concrete)	=600 Pa
25.125 mm Plywood	=165
Framing (estimate)	=120
Ceiling supports	=96
Ceiling	=120
Floor dead load	=1035
+ Partition Load	=**960**
Total Floor Dead Load	**=3096 Pa, say 3100 Pa**

7.1.2 Live Loads

Included in live loads are loads associated with use or occupancy of a structure. Live loads, or imposed loads, are temporary loads of short duration or moving loads. Weights of occupants of a room, stored products, equipment, livestock, furniture, and vehicles are all referred to as floor live loads. The miscellaneous loads that may occur on a roof are referred to as roof live loads. These include loads that are imposed during the construction of the building including the roofing process. Roof live loads that may occur after construction include reroofing operations, air conditioning and mechanical equipment installation and servicing, and, perhaps, loads caused by firefighting equipment. The live load on a roof is usually applied for a relatively short period of time during the life of a structure. This fact is normally of no concern in the design of structures other than wood. However, the length of time for which a load is applied to a wood structure has an effect on the load capacity. The anticipated live loads to be used for building design are specified in the building code that is in force where the building will be constructed. Building codes typically specify the minimum floor life loads and minimum roof life loads that must be used in the design of a structure based on the occupancy or use of the building. Typically, occupancy or use of floor live loads ranges from a minimum of 1915.2 N/m^2 for residential structures to as high as 11,970 N/m^2 for heavy storage facilities.

The structural design for gravity loads involves evaluating each member for the performance under the anticipated live loads, dead loads and a combined force of live load plus dead load often called the total load. The design process starts at the roof and continues down to the foundation. This is opposite to the actual construction which starts at the bottom and works up.

The different forms of dead and live loads include the following:

- **Point Load**: A point load is a concentrated load in kilogramme or Newton at a specific location. This may be the location of bearing of a beam or a post.
- **Uniform Load**: A uniform load is a continuous load along the entire length of a member and is expressed in terms of load per length, e.g., Newton or kilogramme per millimetre.
- **Partial Uniform Load**: A partial uniform load is also expressed in terms of load per length, i.e., Newton or kilogramme per millimetre, but does not run the entire length of the member.
- **Uniform Increasing Load**: Triangular areas are sometimes designed into floor plans and are also sometimes present in residential roofs. Triangular areas can contribute a uniform increasing load to a structural member. Most often an increasing load starts at one end of the member as a zero load and increases to the other end where it is at a maximum load. Structural triangles are usually right triangles. However, complicated or unusual triangular shapes can be solved by trigonometry when encountered.

The majority of wood structural designs are governed by uniform live loads. However, both concentrated and uniform loads should be checked, whichever produces the more critical condition should be used. A load diagram is a working sketch of the loads present on a structural member and is recommended before tackling a complex loading problem. The diagramming convention is to select one end of the member as the left end and locate the loads and their distances towards the right.

7.1.3 Snow Loads

Snow loads represent another type of gravity load that primarily affects roof structures in the temperate countries. Certain types of floor systems including balconies and decks may also be subjected to snow loads.

7.1.4 Lateral Forces

Wind and seismic loads are the two primary lateral forces considered in building design.

Table 7.4 Design wind velocities in Nigeria

Distance of building location from the coast to the hinterland (km)	Condition of exposure of the environment where the building is located	Design velocity (km/h)
0–160	Open country	112
0–160	Built-up areas	96
160–480	Open country	145
160–480	Built-up areas	104
Above 480	Open country	195
Above 480	Built-up areas	128

Source Nigerian Standard Code of Practice NCP (2005)

(a) **Wind Load**: Wind load varies from place to place. They are functions of location, building shape, building height, exposure, importance and gusts. The wind pressures or wind loads on buildings vary with the geographical location and, for a given location, with the height above the ground surface. Winds are assumed to act in an horizontal direction in spite of the direction of prevailing winds. There are positive pressures on the windward side of a building and negative pressures, or side partial vacuums, on the leeward side. Design pressures are based on the total resultant pressure, which is equal to the sum of these pressures. Table 7.4 contains information on design wind velocities in Nigeria.

(b) **Seismic (Earthquake) Load**: Seismic loads are functions of location, importance, building shape, framing system, building mass and foundation soil profile. A number of different forces act on a structure during an earthquake. These forces include inertia forces, damping forces, elastic forces and an equivalent forcing function (mass times ground acceleration). The theoretical solution of the dynamic problem involves the addition of individual responses of a number of modes of vibration. Each mode is described by an equation of motion which includes a term reflecting each of the forces aforementioned. Because earthquake magnitudes are unpredictable, some structural damage is tolerated, but the design attempts to prevent death or injury to occupants.

7.2 Controlling Loads for Design

A load combination results when more than one load type acts on the structure. Design codes usually specify a variety of load combinations together with weighting factors for each load type in order to ensure the safety of the structure under different probable loading scenarios. The reader is advised to refer to Table 3.1 in Chap. 3 where the various load duration factors have been listed.

Example Determine the controlling load duration for a factory roof located in a region where the maximum wind load is 670 Pa. The dead load of the roof structure is 960 Pa, including all equipment, piping and ductwork.

Solution

The amount of load on the structure for these various durations will be determined based on the allowable stresses for the two load durations. Hence, we shall divide the two load combinations by their respective load duration factors of 0.9 for permanent load and 1.33 for wind load as listed in Table 3.1 as follows:

	Dead Load Alone	Dead + Wind Load
	960	960
Wind Load	0	670
Load ÷ Load duration factor	960 ÷ 0.9 = 1067 Pa	1630 ÷ 1.33 = 1226 Pa

The combination of loads that will require the greatest section modulus is Dead Load + Wind Load. Having determined that 1226 Pa for wind load duration controls, one may carry forward with the design using either 1630 Pa and 1.33 × tabulated (allowable) bending stress parallel to the grain of the species or 1226 Pa and tabulated (allowable) bending stress parallel to the grain of the species.

Instances occur where dead load may control if the structure is extremely heavy and carries small or infrequent live loads. Again, the method of simply dividing loads by duration of load factor can be used only if all loads are distributed in the same way, i.e. all uniformly distributed. Dead load is generally uniformly distributed. However, live loads might be differently disposed.

7.3 The Concept of Tributary Loading

Tributary loading is the accumulation of loads that are directed towards a particular structural member. The area that is assumed to load a given member is known as the tributary area. For a beam or girder, this area can be calculated by multiplying the tributary width by the span of the member. The tributary width is generally measured from midway between member(s) on one side of the member under consideration to midway between member(s) on the other side. When the load to a member is uniformly distributed, the load in Newton per millimetre can be readily determined by taking the unit load in N/mm^2 times the tributary width ($N/mm^2 \times mm = N/mm$).

7.4 Structural Design Codes

Structural design codes are legal documents that set forth the minimum requirements to protect the public health, safety and general welfare as they relate to the construction and occupancy of buildings and structures. Most structural codes are periodically revised by professional associations, building officials, industry representatives and other interested parties. The content and administration of design codes vary among countries and regions. Information in this book is based on standard engineering practice and experience with the Nigerian Code of Practice for Timber Designed developed in 1973 and revised in 2005, the Indian Code of Practice for Timber Design, Kenya Bureau of Standards, and to some extent, the American National Design Specification (NDS) for wood construction (1997), revised in 2005 and 2012, respectively.

Worked Example

1. Compute the tributary load for the wooden beam sandwiched between two walls as shown in the figure below if the load is 4788 N/m^2.

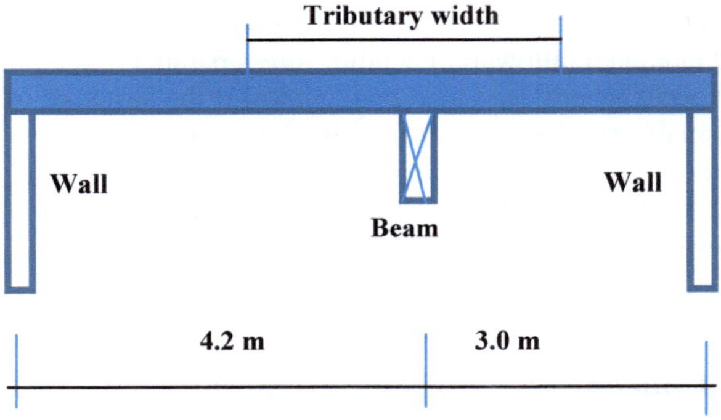

Solution

The load is 4788 N/m^2.

The tributary width of the beam to each outside wall will be half the distance between the outside wall and beam = (4.2/2) + (3/2) = 3.6 m.

Therefore, the tributary load to the beam = 3.6 m × 4788 N/m^2 = 17,237 N/m = 0.0172 N/mm.

Note that the left wall has 2.1 m of tributary width and would receive a load of 2793 N/m. The right wall has 1.5 m of tributary width and gets a load of 1995 N/m.

2. Create an Excel template for computing the dead weight per length (N/mm), weight per area (N/mm^2) and weight per volume (N/mm^3) of structural materials, if the masses (in kg), cross-sectional area (in m^2) and Lengths (in metres) are given.

Solution Steps

1. Open an Excel worksheet.
2. Label the columns as shown in Fig. 7.1a.
3. Input the serial numbers for the materials in the cells of a single column in Excel as shown in Fig. 7.1a.
4. Input the given masses in kg in the cells of the appropriately labelled column.

(a)

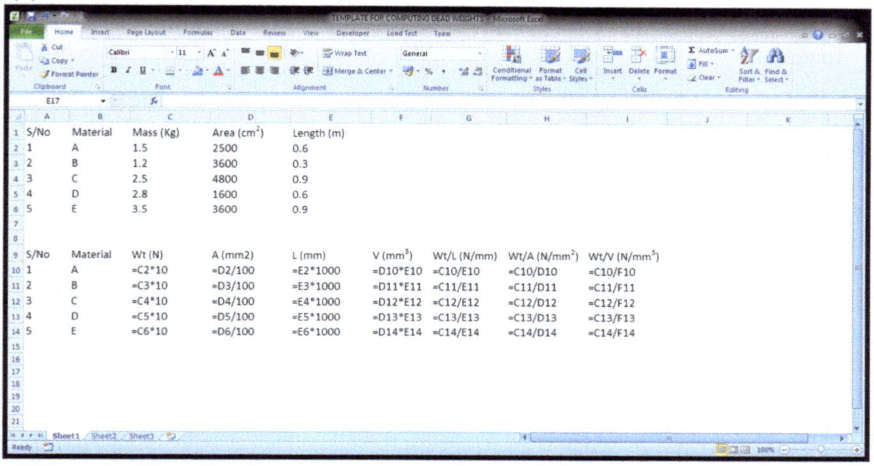

S/No	Material	Mass (Kg)	Area (cm^2)	Length (m)
1	A	1.5	2500	0.6
2	B	1.2	3600	0.3
3	C	2.5	4800	0.9
4	D	2.8	1600	0.6
5	E	3.5	3600	0.9

S/No	Material	Wt (N)	A (mm2)	L (mm)	V (mm^3)	Wt/L (N/mm)	Wt/A (N/mm^2)	Wt/V (N/mm^3)
1	A	=C2*10	=D2/100	=E2*1000	=D10*E10	=C10/E10	=C10/D10	=C10/F10
2	B	=C3*10	=D3/100	=E3*1000	=D11*E11	=C11/E11	=C11/D11	=C11/F11
3	C	=C4*10	=D4/100	=E4*1000	=D12*E12	=C12/E12	=C12/D12	=C12/F12
4	D	=C5*10	=D5/100	=E5*1000	=D13*E13	=C13/E13	=C13/D13	=C13/F13
5	E	=C6*10	=D6/100	=E6*1000	=D14*E14	=C14/E14	=C14/D14	=C14/F14

(b)

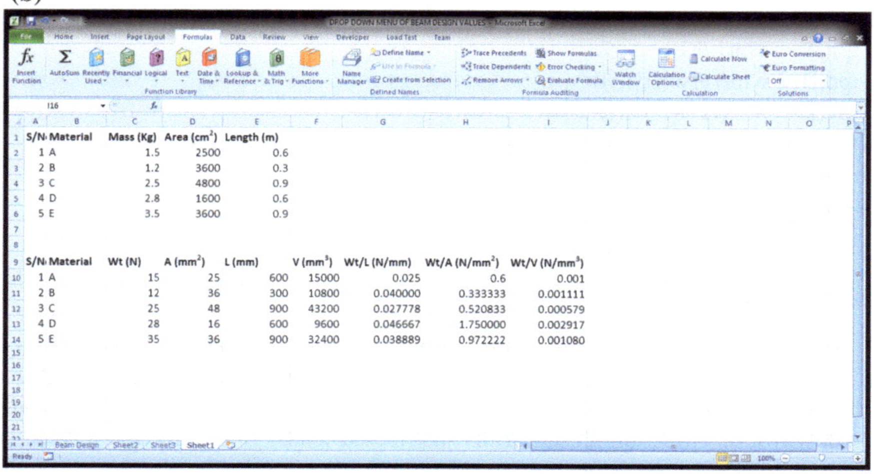

S/No	Material	Mass (Kg)	Area (cm^2)	Length (m)
1	A	1.5	2500	0.6
2	B	1.2	3600	0.3
3	C	2.5	4800	0.9
4	D	2.8	1600	0.6
5	E	3.5	3600	0.9

S/No	Material	Wt (N)	A (mm^2)	L (mm)	V (mm^3)	Wt/L (N/mm)	Wt/A (N/mm^2)	Wt/V (N/mm^3)
1	A	15	25	600	15000	0.025	0.6	0.001
2	B	12	36	300	10800	0.040000	0.333333	0.001111
3	C	25	48	900	43200	0.027778	0.520833	0.000579
4	D	28	16	600	9600	0.046667	1.750000	0.002917
5	E	35	36	900	32400	0.038889	0.972222	0.001080

Fig. 7.1 **a** Excel worksheet showing the computations. **b** Excel worksheet showing the computation results

5. Input the given cross-sectional areas (in kg) and the length (in m) in the cells of appropriately labelled column.
6. Input the formulas for converting the masses (in kg) to weights (in Newton) using the cell references in the appropriate column. Do the same for converting the length (in m) to mm and the area (in m^2) to mm^2 in the appropriate columns.
7. Input the formulas for computing the weight/length, weight/area and weight/volume in the appropriate columns and press 'Enter' each time.

The result of the computation is shown in the Excel worksheet presented in Fig. 7.1b.

3. Write a simple FORTRAN program to compute the dead weight of structural materials.

Solution

```
!Program to compute dead load values of building materials
real::l(100),w(100),t(100),wt(100),wtn(100),wtpl(100),wtpa(100),wtpv(100)
integer::i,j,count
count = 1
write(*,*) 'enter the no of materials'
read(*,*)i
write(*,*) 'enter length(mm), width(mm), thickness(mm) and weigtht(kg)'
do j = 1,i
read(*,*)l(i),w(i),t(i),wt(i)
end do
write(*,*) 's/n,wt/l,wt/a,wt/v'
do j = 1,i
wtn(i) = wt(i)*10
wtpl(i) = wtn(i)/l(i)
wtpa(i) = wtn(i)/(w(i)*t(i))
wtpv(i) = wtn(i)/(w(i)*t(i)*l(i))
write(*,*)count,wtpl(i), wtpa(i),wtpv(i)
count = count + 1
end do
end
```

Practice Questions

1. List and describe briefly five types of loads a structure may be subjected to.
2. A roof structure is to carry live load of 1450 Pa. The dead load will be 725 Pa. The maximum vertical force due to wind pressure will be 865 Pa, acting downward. These are all uniformly distributed loads. Determine the load to be used to design the roof for strength. Show all steps in the derivation of the answer.

References

Goldstein EW (1999) Timber construction for architects and builders. McGraw-Hill Publishers, New York

Gurfinkel G (1973) Wood engineering. Southern Forest Products Association, USA

Lindley JA, Whitaker JH (1996) Agricultural buildings and structures, revised edn. The American Society of Agricultural Engineers, USA

Midwest Plan Service (1977) Structures and environment handbook. Midwest Plan Service Ames, Iowa

NCP (2005) Nigerian code of practice for structural timber design. The Standards Organisation of Nigeria

Ozelton EC, Baird JA (1976) Timber designers' manual. Granada Publishing Limited, London

Seward D (2003) Understanding structures, analysis, materials, design, 3rd edn. Palgrave Macmillan, New York

Shaeffer RE (1999) Elementary structures for architects and builders, 4th edn. Prentice-Hall Publishing Co., New Jersey

Willenbrock JH, Manbeck HB, Suchar MG (1998) Residential building design and construction. Prentice Hall, New Jersey

Chapter 8
Design of Wooden Beams

8.1 Stresses in Beams

A beam is a structural member that carries loads perpendicular to the longitudinal axis. Either forces or couples that lie in a plane containing the longitudinal axis of the beam may cut upon the member. If the forces act perpendicular to the longitudinal axis, the plane containing the forces is assumed to be a plane of symmetry of the beam. These forces and couples tend to:

- Impart deflections perpendicular to the longitudinal axis of the bar, and
- Set up both normal and shearing stresses on any cross-section of the beam perpendicular to its axis.

A beam undergoes ***pure bending*** when only couples are applied to its ends and no forces act on the bar. A beam subjected to pure bending has only normal stresses with no shearing stresses set up in it. Bending produced by forces that do not form couples is referred to as ***ordinary bending***. A beam subjected to ordinary bending has both normal and shearing stresses acting within it.

It is usually assumed that a beam is composed of an infinite number of fibres, moving independently of every other fibre, i.e. no lateral pressures or shearing stresses between the fibres.

Figure 8.1 depicts a board supported at each end on its narrow face and bending under a load, i.e. acting as an elastic beam. The application of the load causes downward deflection of the beam and bending stresses. The fibres in the upper edge of the beam are resisting compression while those on the lower edge are experiencing tensile stress, i.e. the lower part of the beam is resisting separation (the strain will tend to extension in beam length) while the upper part is experiencing compressive stress (the strain will tend to contraction in length), and there is shear stress

A.O. Olorunnisola, *Design of Structural Elements with Tropical Hardwoods*,
DOI 10.1007/978-3-319-65343-3_8

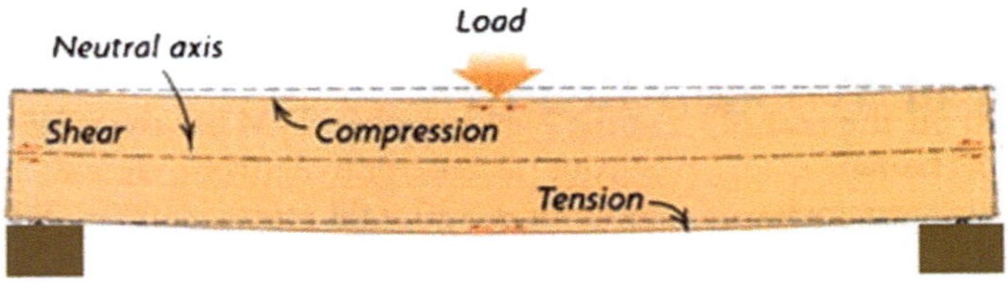

Fig. 8.1 Bending in a simply supported beam

around a neutral axis—where length does not change—but the axis moves lower as the load increases. Simply put, all fibres below the neutral axis are in a state of tension, while those above the neutral axis are in compression. The neutral axis always passes through the centroid of the cross-section. The distribution of the tensile, compressive and shear stresses has implications in the design of beams, leading to the use of more complex, but lighter, beams in place of solid beams as will be discussed shortly.

8.2 Bending Equation

For any beam having a longitudinal plane of symmetry and subject to a bending moment (M) at a certain section, the normal stress activity on a longitudinal fibre at a distance (y) from the neutral axis of the beam is given by:

$$\sigma = \frac{\mathrm{My}}{I} \quad \text{or} \quad \sigma = \frac{\mathrm{M}}{I/y}$$

where I = second moment of inertia of the cross-sectional area about the neutral axis. The ratioI/y is called the section modulus (S), whose units are mm^3 or cm^3. The maximum bending stress in a beam can, therefore, be represented by the following equation:

$$\sigma = \mathrm{M/S}$$

In the derivation of the bending stress equation it is assumed that:

(i) Transverse sections of the beam (i.e. plane sections of the beam normal to its longitudinal axis) that were plane before bending remain so even after bending.
(ii) The material of the beam is isotropic and homogeneous and follows Hooke's law.
(iii) The beam is initially straight and of constant cross-section.
(iv) The moduli of elasticity for tension and compression are equal.

(v) The stresses in the beam do not exceed elastic limit.
(vi) The beam is subjected to pure bending (only bending moment, no sharing) and therefore bends in an arc of a circle.
(vii) The radius of curvature is large compared to the dimensions of the section.
(viii) The plane of loading contains a principal axis of the beam cross-section and the loads act perpendicular to the beam axis.
(ix) The cross-section of the beam does not change abruptly.

Some of these assumptions are not valid for wooden beams. For example, wood is anisotropic and not isotropic as stated in the second assumption. Nevertheless, the bending stress equation has been adopted for the design of wooden beams. A very useful practical implication of the equation is that the bending stress in a given beam at a particular point is proportional to the distance of that point from the neutral axis (y). The maximum bending stress occurs when y is maximum (at the outermost surfaces). The bending stress at the neutral axis (where $y = 0$) is zero. As such, it is economically desirably to so design a beam as to provide maximum possible sectional area as far away as possible from the neutral axis. This explains the use of I-sections (Fig. 8.2), and the built-up sections as wooden beams.

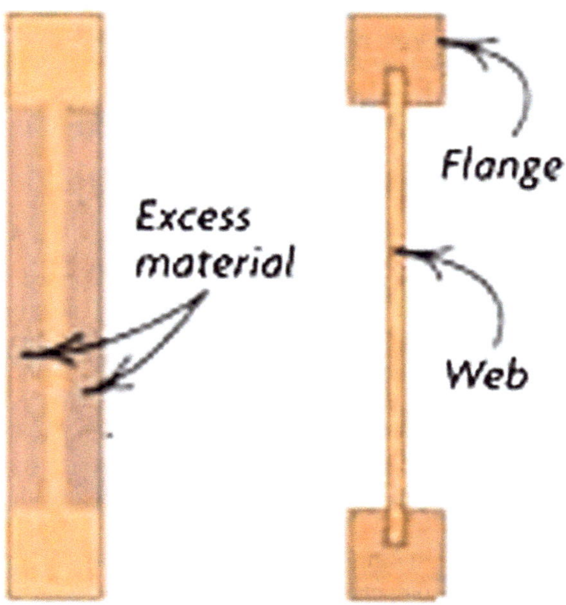

Fig. 8.2 An I-beam

8.3 General Design Information for Designing Wooden Beams

In designing a wooden beam, the designer needs to have, ab initio, or generate the following items of information:

(i) **Design Loads**: Design load consists of the dead load, i.e. the weight of the structure plus any permanently fixed loads and the live loads. The live load may be taken from a code or a design standard, or it may be determined by experience with the intended use of the structure.

(ii) **Span**: The effective span of a beam as a structural member is the horizontal distance from face-to-face supports, plus half (½) of the required length of bearing at each end. For continuous beams, the span length is measured from the centre of the bearing at those supports over which the beam is continuous.

(iii) **Design Values**: Unit design values for the design of wooden beams, i.e. bending strength parallel to grain, shear strength parallel to grain, modulus of elasticity for determining maximum deflection and compression strength perpendicular to grain for determining bearing strength, are given in the national design codes of various countries and are subject to appropriate modification/ adjustment factors for duration of load and other conditions of use.

(iv) **Net Sizes of Lumber**: Lumber is customarily specified in terms of nominal sizes. Computations used in design should preferably be based on the net dimensions, i.e. actual sizes. If this is not possible, then a factor of safety should be applied to adjust the expected load.

8.4 Actual Wooden Beam Design

The basic objective in structural analysis and design of a beam is to produce a structural element capable of resisting all applied loads perpendicular to its long axis without failure during its intended life. The primary purpose of a structure is to support live and dead loads. If the beam is improperly designed or fabricated, or if the actual applied loads exceed the design specifications, the beam will probably fail to perform its intended function, with possible serious consequences. A well-designed beam greatly minimizes the possibility of costly failures.

The design of wooden beams, just like any structural element, requires many detailed computations, some of which are of a routine nature. The wider availability of the computer in the last two decades has resulted in its rapid adoption for structural design in replacement for manual computation. Thus, few examples on the use of computers for the design of wooden beams are presented . The reader should note that, as earlier mentioned, many countries have their own structural design codes and that it is necessary for a designer to become familiar with local requirements or recommendations in regard to correct practice. The examples given in this chapter are meant to demonstrate design procedures. They should not be assumed to apply to all areas or situations.

Investigation of the strength and stiffness requirements of a wooden beam under transverse loading should take into consideration of the following factors:

(i) Bending moment/stress induced by the load.
(ii) Horizontal shear stress at supports.
(iii) Deflection or deformation caused by the load.

Any one of these three factors may control the design although deflection is not a matter of safety and would be a control only where appearance or comfort of the occupants is important. Designing a wooden beam or selecting member sizes of a wooden beam therefore consists of:

- Providing enough section modulus so that the allowable bending stress is not exceeded.
- Providing enough cross-sectional area so that the allowable shearing stress is not exceeded.
- Providing enough moment of inertia so that the permissible deflection is not exceeded.

In most wooden beam problems, the member is loaded about the strong axis of the cross-section. Therefore, the width of a beam is usually the smaller cross-sectional dimension, and the depth is the larger. Naturally, the strong axis has larger values of section modulus and moment of inertia. Loading a beam about the strong axis (x-axis) is also described as having the load applied to the narrow face of the beam. If the bending stress is about the weak axis (y-axis), the section modulus ($bd^3/12$) and moment of inertia ($bd^2/6$) are much smaller. Decking is an obvious application where a beam will have the load applied to the wide face of the member. In this case, the width is the larger cross-sectional dimension, and the depth is the smaller. This type of beam loading is less common. More often than not, a wooden beam is used in bending about the strong axis whenever possible.

8.4.1 Bending Stress Criterion

To maintain static equilibrium in a beam, the bending strength of the member must not be less than the bending stress induced by the live and dead loads on the beam, i.e.

$$\text{Bending Strength of the wood} \geq \text{Induced Bending Stress in the beam, i.e.,}$$
$$\boldsymbol{\sigma}_{\text{b}} \times (\text{applicable modification factors}) \geq \boldsymbol{M/S}$$

where

σ_b = Tabulated bending strength of the selected wood species (with specific modification factors applied with reference to the code of practice adopted for the design).
M = Maximum bending moment induced in the beam by the load (based on the type of loading and type of support).
S = Section modulus of the beam.

Moment: The more common loading condition is to stress a wooden beam about its strong axis, also known here as the x-axis, taking advantage of the larger moment of inertia that works to reduce bending stresses. This is the condition shown in Fig. 8.3.

The maximum induced moments in single span beams under different types of loading are presented in Table 8.1.

Section Modulus: The section moduli (S_x) for rectangular, square and circular cross-sections are:

$$S_x = \frac{bd^2}{6} \quad \text{for a rectangular cross-section}$$
$$S_x = d^3/6 \quad \text{for a square cross-section}$$
$$S_x = \pi d^3/32 \quad \text{for a circular cross-section}$$

As already noted, the bending strength must be equal to or greater than the induced bending stress in order to maintain static equilibrium.

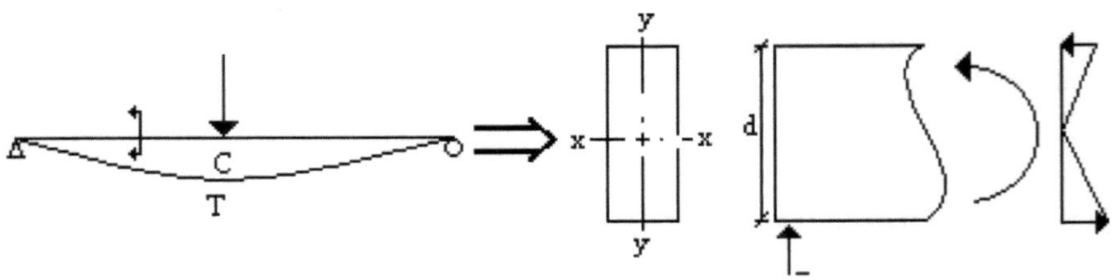

Fig. 8.3 Strong axis bending due to edgewise loading

Table 8.1 Maximum moment, shear force and deflection in beams under different loading conditions

Loading condition	Maximum moment (M)	Maximum shear force (V)	Maximum deflection (Δ_{max})
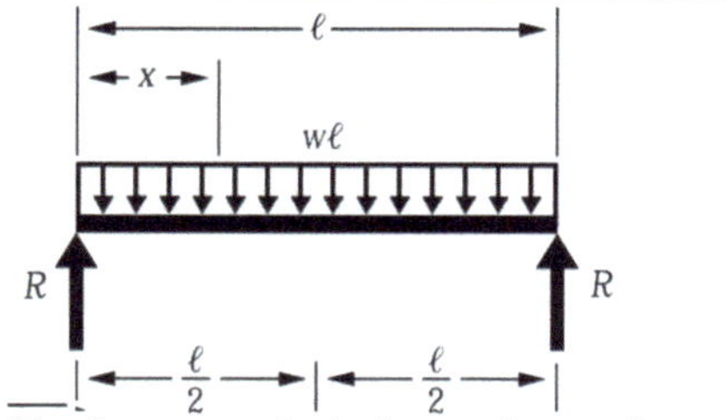 Simply supported, single span beam of span L, under uniform loading (w)	$\frac{wL^2}{8}$	$\frac{wL}{2}$	$\frac{5wL^4}{384EI}$
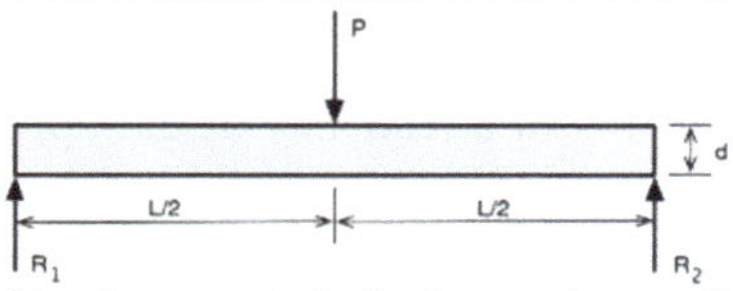 Simply supported, single span beam of span L, under concentrated load (P) at mid-span	$\frac{PL}{4}$	$\frac{P}{2}$	$\frac{PL^3}{48EI}$
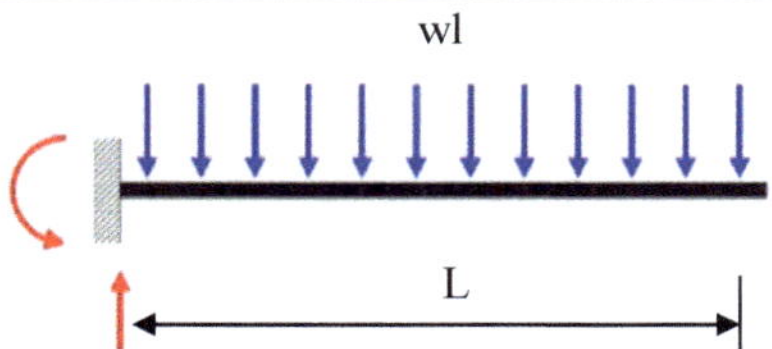 Cantilevered single span beam of span L, under uniform loading (w)	$\frac{wL^2}{2}$	wL	$\frac{wL^4}{8EI}$
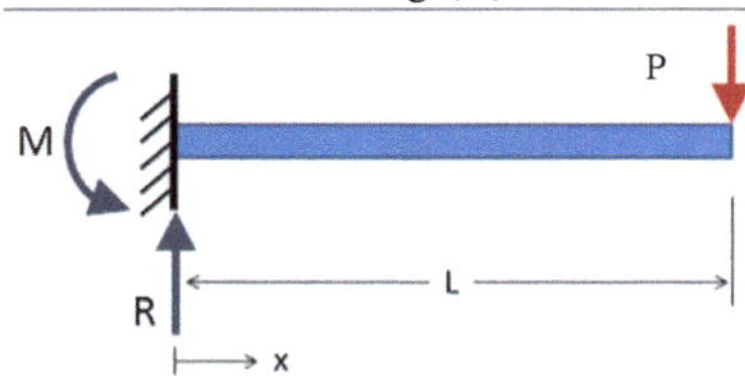 Cantilevered single span beam of span L, under concentrated load at the free end	PL	P	$\frac{PL^3}{3EI}$

L = span of the beam (mm)
w = uniformly distributed load on the beam (N/mm)
P = concentrated load on the beam (N)
d = thickness (narrow face) of the beam (mm)
b = width (wider face) of the beam (mm)
I = moment of inertia (mm^4)
E = modulus of elasticity (N/mm^2)

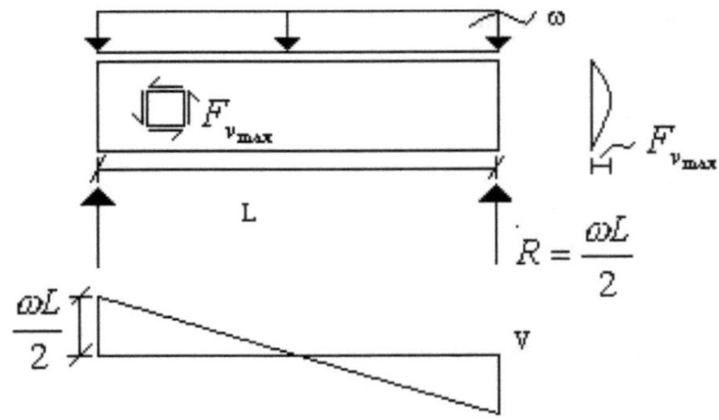

Fig. 8.4 Shear stress profile in a beam subjected to uniformly distributed load

8.4.2 Shear Stress Criterion

A beam subjected to a vertical shearing force is also subjected to a horizontal or longitudinal shearing load. Such a vertical load results in a tendency of the upper part of the beam to slide by the lower part as shown in Fig. 8.4.

To maintain equilibrium within the beam, the shear resistance of the wood must be equal to or exceed the horizontal shear induced by the vertical load, i.e.

$$\text{Shear Strength of the wood} \geq \text{Induced Shear stress in the beam}$$
$$F_v \times (\text{applicable modification factors}) \geq f_v$$

This means that for design purposes, the calculated f_v may not exceed the tabulated horizontal shear stress (also referred to as the design value in horizontal shear), F_v, for the species and grade of lumber used. The maximum intensity of horizontal shear stress occurs at the neutral axis of the section and is dependent on the magnitude of the vertical shear force. The maximum horizontal shear stress in a wooden beam is obtained from the formula:

$$f_v = \frac{VQ}{Ib}$$

where

V = Maximum shear force (N)
Q = Statical moment of area about the neutral axis (mm^3)
I = Moment of inertia, (mm^4)
b = breadth of rectangular beam, (mm).

$$\text{For a rectangular or square beam, } f_v = \frac{1.5\,V}{A}$$

$$\text{For a circular beam, } f_v = \frac{1.33\,V}{A}$$

where

A = Area of the cross-section (mm^2).

The maximum shear force, V, to be inserted in any of the shear stress equations given above depends on the type of loading on the beam as shown in Table 8.1.

Notching of beams, i.e. cutting away some portion of the beam at the two extreme ends often done so that that it can rest properly on the supports as shown in Fig. 8.5, should be avoided especially on the tension side of the member. Notches at the ends do not affect bending strength directly, but do affect shear strength. Generally, a number of design codes specify that notches in sawn lumber beams shall not exceed 1/6th of the depth of the member and shall not be located in the middle third of the span. Where members are notched at the ends the notched depth shall not exceed one-fourth the beam depth. The tension side of sawn lumber beams of 100 mm (4 inches) or greater nominal thickness shall not be notched except at ends of members.

The Nigerian Code of Practice for Timber Design (NCP 2005) specifies that when designing a beam having square—cornered notches at the ends as shown in Fig. 8.6, the tabulated shear strength should be multiplied by the modification factor:

$$F_4 = \frac{\text{effective depth}}{\text{Total depth}} = \frac{d_e}{d}$$

i.e.

design shear strength = $F_v \cdot F_4$

where

F_v = Shear strength of the wood material.

Fig. 8.5 A design arrangement incorporating a notched wooden beam

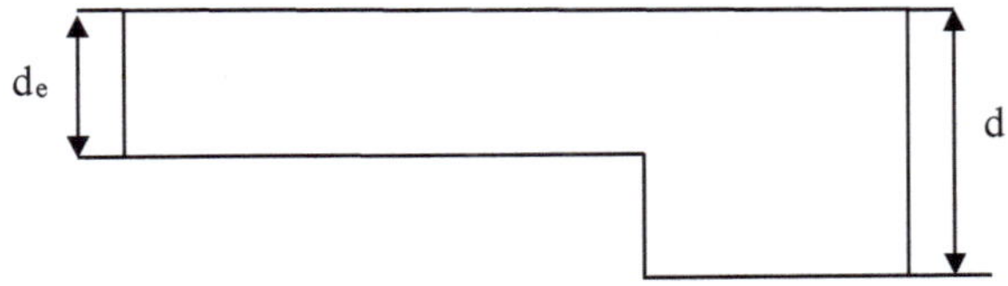

Fig. 8.6 Schematic diagram of a notched beam

8.4.3 *Deflection Criterion*

The deflection of a beam (Fig. 8.7) is a measure of the deformation that occurs as the beam resists bending under applied load. When the induced bending stress does not exceed the applicable design value, this deformation does not seriously affect the endurance of the beam. Thus, where appearance or rigidity of the assembly is not important, deflection may be ignored and the design based on strength alone. However, where appearance or rigidity is important, deflection may be the controlling factor in determining the size of member required. Another reason for limiting deflection is to control vibration due to impact on residential floors.

The reason for controlling deflection again has a bearing on the design load selected. Deflection due to the dead load of the materials of construction would have occurred by the time they are installed. Where the purpose is to provide adequate rigidity to avoid damaging brittle materials or to eliminate excessive vibration in floors due to impact, design for live load only is adequate.

The deflection of a wooden beam, under long-continued, full design load, will increase beyond what it was immediately after the load was first applied but without endangering the safety of the beam. Where it is necessary to limit the deflection under such long-term loading, extra stiffness can be provided in the design stage by increasing member size. This can be done by applying an increase factor to the deflection due to long-term load. Total deflection is thus calculated as the immediate deflection due to long-term or permanent loading, multiplied by the appropriate deflection factor plus deflection due to the short-term component of the design load. It is customary practice to use a deflection factor of 1.5 for

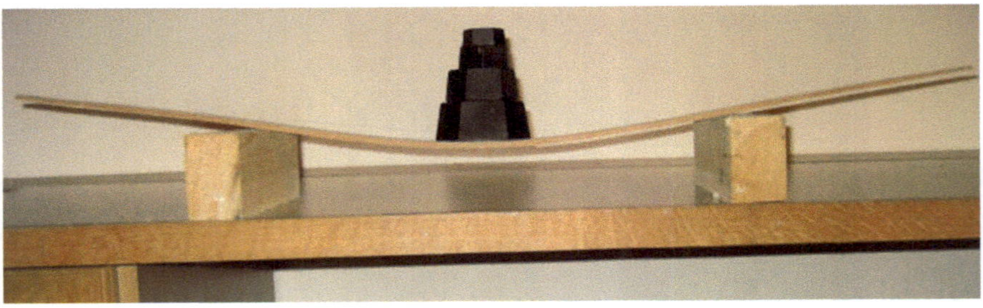

Fig. 8.7 Deflection in a wooden beam

glued-laminated timber or seasoned sawn lumber, or 2.0 for unseasoned sawn lumber, when calculating deflection due to long-term loading, i.e.

Long Term, Total Δ Estimates for seasoned lumber or glulams:

$$\Delta_{total} = \underbrace{(1.5\Delta_{DL+\%LL})}_{\text{Long Term}} + \underbrace{\Delta_{LL}}_{\text{Short Term}}$$

In any case, it should be understood that the recommended values for modulus of elasticity will give the initial deflection of a beam and that this will increase under long-continued, full design load. The governing equation for deflection is:

Allowable deflection (deflection limit) ≥ Maximum induced deflection, i.e.,

$$\Delta_{all} \geq \Delta_{max}$$

Deflection limits (Δ_{all}) are usually expressed as a fraction of the span. The selection of an appropriate limit has generally been a matter of judgment on the part of the designer. The most generally used deflection limits for a beam of span L are:

Floor beams

- $\Delta_{all} = L/360$ (mm) for floor joists carrying live load only
- $\Delta_{all} = L/240$ (mm) for floor joists carrying Dead + Live load.

Roof beams

- $\Delta_{all} = L/240$ (mm) for roof framing with slope of 3 in 12 or less or roof bean carrying live load only:
- $\Delta_{all} = L/180$ (mm) for roof framing with slope of more than 3 in 12 or roof beam carrying Dead + live load
- $\Delta_{all} = L/360$ (mm) for ceiling joists.

The maximum deflection, i.e. the resistance to deflection of a beam under load (Δ_{max}), is provided by the product of the modulus of elasticity (E) and moment of inertia (I) and is dependent on the type of beam support and loading conditions as shown in Table 8.1. The tabulated deflection formulas only take into account deflection due to bending stress and do not take into account deflection due to horizontal shear in planes parallel to the neutral axis, which is sometimes added. For example, the shear deflection in wooden beams subjected to uniformly distributed load is computed thus:

$$\Delta_s = \frac{KwL^2}{8AG}$$

where

Δ_s = Shear deflection (mm)
w = uniformly distributed load (N/mm)
L = Span of the beam (mm)

G = Modulus of rigidity or shear modulus of the wood member (N/mm^2)
A = Cross-sectional area of the beam (mm^2)
K = A constant depending on the shape of the beam (1.2 for rectangular beams and 1.1 for round beams).

Hence, the total deflection in a wooden beam subjected to uniformly distributed load is the sum total of the deflection due to both bending (Δ_b) and shear (Δ_s), i.e.

$$\Delta_{\text{Total}} = \Delta_b + \Delta_s$$

$$\Delta_{\text{Total}} = \frac{5wL^4}{384EI} + \frac{KwL^2}{8AG}$$

Hence, the governing equation for deflection now becomes:

$$\text{Allowable deflection (deflection limit)} \geq \text{Total induced deflection, i.e.,}$$
$$\Delta_{\text{all}} \geq \Delta_{\text{Total}}$$

For a specified deflection limit, the appropriate formula may be used to determine any one of the four factors of E, I, W or L provided the other three factors are known. The preceding formulas apply only for uniformly distributed load or concentrated load at mid-span on simple supports at both ends. Formulas for other conditions of loading may be obtained from textbooks on strength of materials.

It is important to note that bending stress usually controls member size except for short, heavily loaded spans and overhangs, where shear can govern. The span range for dimension lumber which is nominally 50–100 mm (2″–4″) extends to approximately 6 m (20ft), and deflection will often control the design when the span is this great. Glued-laminated sections, plywood box beams, and large timbers can span much farther. Also, a wood structure's self-weight is seldom a design factor except for large members. One method of selecting wooden beams is to size the beam to meet the requirements for moment and then check to see that the area provided is sufficient for shear and that the moment of inertia provided is enough to control the deflection. For a heavily loaded short span, where shear is likely to control, an alternative procedure of sizing for shear and then checking for moment and deflection might save time.

Any increase in span of a beam almost always increases load by gathering more tributary area for the member, so the effects of load and span are difficult to differentiate. Assuming, however, that change in span does not necessarily mean change in load, the following can be stated with respect to simple wooden beams:

- Increasing the span alone will cause a proportional change in moment, no change in shear and increase deflection as the cube of the relative span change.
- Increasing a load while making no other changes will cause a proportional increase in moment, shear and deflection.
- Changing a uniform load to a concentrated one of the same magnitude will cause the moment to double, no change in shear and a deflection increase of approximately 50%.

Also, the breaking strength of a beam is directly proportional to the width, but proportional to the depth squared and inversely proportional to the span. The stiffness is directly proportional to the width, proportional to the depth cubed and inversely proportional to the span cubed. Thus, whether the depth or width of a solid beam is doubled the mass is the same but the strength/stiffness/weight ratios for a beam are very much better with increased depth rather than increased thickness.

8.4.4 *Typical Cases of Wooden Beam Design*

The governing equations for some of the most common wooden beam loading and support conditions may be written as follows, depending on what is required:

Case I: To determine the beam size required for a given span and load

(i) Select a trial cross-section (beam size).
(ii) Compute the maximum bending moment (M) and section modulus (S).
(iii) Compute the induced bending stress, i.e. maximum bending moment (M) divided by the section modulus (S).
(iv) Compare the bending strength of the wood with the induced bending stress.

If the bending strength is greater than the bending stress, then the trial section is Ok and the beam will not fail in bending. Otherwise, select a **larger** beam size. Typically, the minimum beam size you should select should be 50 × 100 mm (2″ × 4″). However, be careful not to overdesign, i.e. do not select a member that is way too big for the load it is expected to carry since this has cost implications.

(v) Compare the shear strength of the wood with the induced shear stress . If the shear strength is greater than the shear stress, then the trial section is still Ok and the beam will not fail in shear. Otherwise, discard the initial beam dimensions and select a **larger** beam size.
(vi) Compare deflection limit with the induced deflection. If the deflection limit is greater than the induced deflection, then the trial section is still Ok and the beam will not fail in shear. Otherwise, discard the initial beam dimensions and select a **larger** beam size.

Case II: To determine the allowable span for a given size of beam and load

(i) Select a trial span (beam length).
(ii) Compute the induced bending stress.
(iii) Compare the bending strength of the wood with the induced bending stress.
(iv) If the bending strength is greater than the bending stress, then the span is Ok and the beam will not fail in bending. Otherwise, select a **smaller** span.

(v) Compare the shear strength of the wood with the induced shear stress. If the shear strength is greater than the shear stress, then the trial span is still Ok and the beam will not fail in shear. Otherwise, discard the initial beam dimensions and select a **smaller** span.

(vi) Compare deflection limit with the induced deflection. If the deflection limit is greater than the induced deflection, then the trial span is still Ok and the beam will not fail in shear. Otherwise, discard the initial trial span and select a **smaller** span.

Case III: To determine the allowable load for a given span and size of beam

(i) Select a trial load.

(ii) Compute the induced bending stress.

(iii) Compare the bending strength of the wood with the induced bending stress.

(iv) If the bending strength is greater than the bending stress, then the load is Ok and the beam will not fail in bending. Otherwise, select a **smaller** load.

(v) Compare the shear strength of the wood with the induced shear stress. If the shear strength is greater than the shear stress, then the trial load is still Ok and the beam will not fail in shear. Otherwise, discard the initial trial load and select a **smaller** load.

(vi) Compare deflection limit with the induced deflection. If the deflection limit is greater than the induced deflection, then the trial load is still Ok and the beam will not fail in shear. Otherwise, discard the initial trial load and select a **smaller** load.

8.4.5 Design of Bearing Supports

The load on a wooden beam tends to compress the wood fibres at points where the beam rests on supporting members as shown in Fig. 8.8a. Thus, the area of bearing on such supports must be large enough to prevent damage to the wood fibres. Such required bearing area is determined by dividing the reaction force by the tabulated compression strength perpendicular to grain for the species and grade of lumber to be used, i.e. the governing equation for the design of bearing area is:

$$\text{Compression strength perpendicular to the grain of the wood} \geq \frac{\text{Support reaction force}}{\text{Bearing Area}}$$

$$\sigma_{\text{cperp}} \geq \frac{P}{w \times l_{\text{b}}}$$

where

w = bearing width = width or diameter of the beam
l_{b} = bearing length as shown in Fig. 8.9.

Fig. 8.8 A wooden bearing support for a set of wooden beams

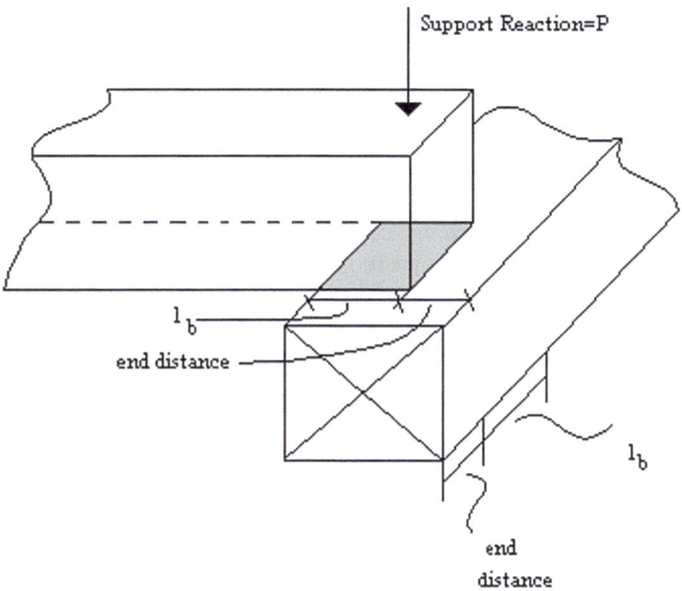

Fig. 8.9 Bearing support dimensions

The Nigerian Code of Timber Design (2005) stipulates that for bearings less than 150 mm (6 inches) long, located 75 mm (3 inches) or more away from the end of a wooden beam, higher compression strength perpendicular to the grain may be used

Table 8.2 Modification factors for different bearing lengths

Length of bearing (mm)	12	25	40	50	75	100	125 or more
F_2	1.70	1.53	1.34	1.19	1.14	1.10	1.00

safely, i.e. the design value in compression perpendicular to the grain may be multiplied by the factor F_2 given in Table 8.2.

8.4.6 Check for Lateral Stability

Rectangular beams which are relatively deep in comparison with their width may be unstable under the application of loads. Such instability is due to the tendency of the compression edge of the beam to buckle causing the beam to deflect laterally. The following general rules have been stipulated by the Nigerian Code of Timber Design (2005) in providing internal restraint for sawn lumber rectangular beams:

If the ratio of depth-to-breadth (d/b) based on nominal dimension is:

(a) 2:1: no lateral support shall be required.
(b) 3:1 or 4:1: the ends shall be held in position, as by full depth solid blocking, bridging, hangers, nailing or bolting to other members or other acceptable means.
(c) 4:1: the ends shall be held in position and member held in line, as by purlins or tie rods.
(d) 5:1: the ends shall be held in position and compression edge held in line, as by direct connection of sheathing, deck or joist.
(e) 6:1: the ends shall be held in position and compression edge held in line, as by direct connection of sheathing, deck or joist together with adequate bridging or blocking spaced at intervals not exceeding 6 times the depth.
(f) 7:1: the ends shall be held in position and both edges shall be held in line.

A more precise method of beam design which accounts for internal stability involves computing the slenderness factor, R_B, and using it in categorizing a particular beam as short, intermediate or long.

$$R_B = \sqrt{L_e d / b^2}$$

where

L_e = effective length of the beam (mm), usually computed based on formulas obtained from timber design codes
d = depth of the beam (mm)
b = breadth of the beam (mm).

If $R_B \leq 10$, the beam is said to be short and the tabulated (full) allowable bending stress adjusted for various conditions is used in designing the beam.

If $10 < R_B \leq R_k$, the beam is said to be intermediate and the allowable bending stress is computed as follows:

$$\sigma_B^1 = \sigma_B\left[1 - 0.33(R_B/R_k)^4\right]$$

where $R_k = 0.775\sqrt{E}/\sigma_B$.

If $R_k < R_B \leq 50$, the beam is said to be long and the allowable stress is computed as follows:

$$\sigma^1 = 0.40E/R_B^2$$

8.4.7 Design for Combined Bending and Axial Loading

Loading conditions on a beam are sometimes of a nature which induces bending and axial tension or compression in the member at the same time. When this condition is expected to exist, the member must be designed to resist the combined forces without exceeding the allowable unit stresses. In the case of combined bending and axial compression, the induced compression stress parallel to grain is given as $F_c = P/A$ while the induced bending stress equation remains $\sigma_b = M/S$. The member is in equilibrium when:

$$\frac{F_c}{\sigma_c} + \frac{F_b}{\sigma_b} \leq 0.9$$

where

P = The compressive force acting on the beam.
A = Cross-sectional area of the beam.
σ_c = Compression strength parallel to the grain of the wood multiplied by all applicable modification factors.
σ_b = Bending strength of the wood multiplied by all applicable modification factors except beam stability factor.

8.4.8 A Summary of the Procedures for Designing a Wooden Beam

1. Sketch the beam based on the information available/given.
2. State the available/given design parameters and values tabulated in the design code.

3. Select a trial cross-section or span or load, depending on the given design question.
4. Compute the following parameters: the maximum bending moment (M), the maximum shear force (*V*) and shear stress (f_V) on the beam and the allowable deflection (Δ_{all})
5. Use the beam design governing equations, applying the appropriate modification factors, to check for compliance with bending stress, shear stress, deflection and bearing stress criteria, i.e.
   - $\sigma_b \times$ (applicable modification factors) $\geq$ M/S
   - $F_v \times$ (applicable modification factor) $\geq f_v$
   - $\Delta_{all} \geq \Delta_{max}$
   - $\sigma_{cperp} \times$ (applicable modification factor) $\geq \frac{P}{A_{bearing}}$
6. Check for lateral stability of the beam is rectangular.

8.5 Special Wooden Beams

There are different types of wooden beams that do not fit the description of a solid wooden beam. Some of these types of wooden beams are briefly introduced in this section.

8.5.1 Box Beams

There are times when solid wooden beams are not suitable for a particular purpose. At such times, wooden box beams may come under consideration. Wooden box beams are created using sections of lumber as shown in Fig. 8.10a, or sometimes along with plywood sheets as shown in Fig. 8.10b. Unlike a solid beam, a box beam is hollow inside. Creating a box beam with lumber sections is a relatively simple task. Four lengths of lumber may be joined together to create the box. The four sides of the box may be secured in place with nails and/or wood adhesive. Once the beam is assembled, it is ready for use as a support for a porch roof or in a number of other practical and decorative applications. In the case of lumber–plywood box beams, the lumber serves as the flanges which carry most of the bending moment, while the plywood is used for webs to carry the shear stresses . The plywood and lumber are usually joined together with adhesives to obtain sufficient pressure for good mating of the surfaces. Vertical lumber stiffeners are often placed between the flanges to prevent web buckling and to transfer load and bearing forces between the flanges.

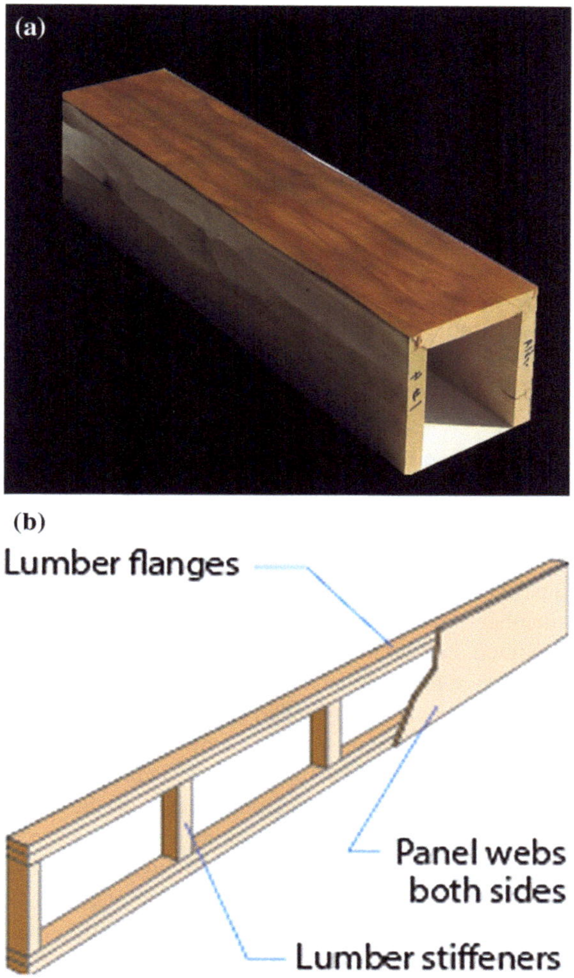

Fig. 8.10 **a** A lumber box beam. **b** A typical plywood–lumber beam

While it may not be as sturdy as a solid wooden beam, the box beam does have several practical advantages. A major one is its relatively light weight which often makes it easier to move into position. Another advantage is that box beams can be made to almost any imaginable size, i.e. they can be as long as 18 m (60 feet) and as large as 750 × 750 mm (30 × 30 inches). In some cases, the hollow space inside the box beam can be used to conceal steel, wiring, ductwork, etc., while still keeping the look of a solid timber.

Box beams are usually designed to job-specific standards and are typically assembled off-site. They are especially suitable in areas that require a long span such as kitchens, entertainment halls, dens, sports arenas, commercial offices, and warehouses since they are lightweight and can be used as a decorative feature. The design procedure for a typical lumber–plywood beam is as follows:

- Estimate the approximate size of the beam required, i.e. select a trial section in terms of the beam depth, plywood thickness and the size of the lumber for the flanges. It is usually safe to assume that all of the bending moment is resisted by the lumber flanges while the shear is resisted by the plywood webs. Proportions of depth-to-span of 1/8–1/12 have been found suitable for many designs.
- Determine the bending moments, moment of inertia and section modulus.
- Determine the horizontal shear and required plywood thickness.
- Determine the shear at the flange–web joint, and choose web and flange dimensions and arrangements to provide adequate rolling shear area.
- Calculate the deflection based on the section properties and compare to the allowable deflection.
- Determine the bearing stiffener size and spacing.
- Determine the splice details for webs and flanges.

8.5.2 Composite Beams

The term composite beam applies to members composed of two or more smaller members joined by nailing, bolting or other mechanical fastening. Structural members can be built up too desired size or shape with small pieces of lumber if solid-sawn members are not available or practical. The individual members must be fastened together and loaded in such a way that the built-up beam acts as a solid unit. Wooden beams are also sometimes reinforced with steel plates as shown in Fig. 8.11a to make them stronger. Such beams are known as composite or flitch beams. The steel plates must be located on the sides of the wood beam so that they are in bearing at the support points. Steel has a modulus of elasticity of about 20 times that of the most commonly used structural timber. This translates to a potential increase in the strength/volume ratio of about 20, as well as stronger and more compact joints with less timber by reinforcing highly stressed areas with light gauge steel. The reinforcing plate may or may not be of the same width as the beam and is usually screwed or bolted to the beam. Since the two materials are rigidly connected, it is assumed that the strains due to bending stress are the same in both materials.

Hence for a flitch beam comprising a wooden beam of breadth ‘*b*’ and depth ‘*d*’, reinforced with two steel plates, each of thickness ‘*t*’ and depth ‘*d*’ (as shown in Fig. 8.11b), if E_w and E_s represent the moduli of elasticity, while σ_w and σ_s represent the maximum stresses developed in wood and steel, respectively, the strains in the two materials are equal, i.e.

$$\sigma_w/E_w = \sigma_s/E_s$$

$$\sigma_s/\sigma_w = E_s/E_w = \text{m(commonly referred to as modular ratio)}$$

$$\text{Therefore, } \sigma_s = \text{m}\sigma_w$$

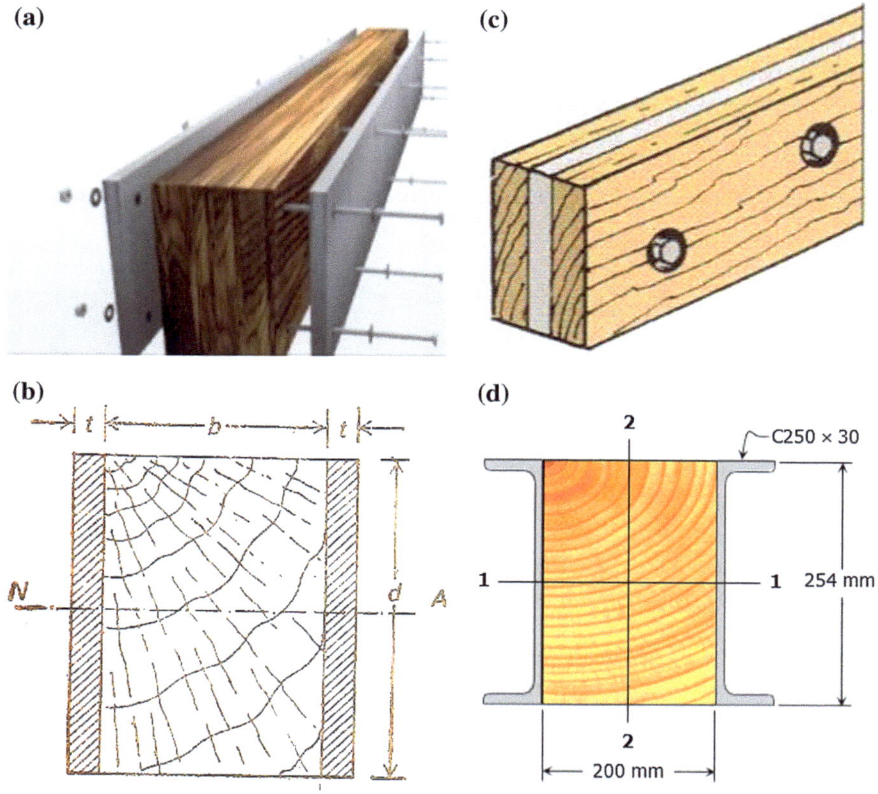

Fig. 8.11 **a** A flitch beam. **b** Schematic diagram of a composite beam. **c** A steel plate between two wood beams. **d** Two angles fitted around the edge of the wood beam

If M_w and M_s represent the moments of resistance of the wooden beam and the reinforcing steel plates, respectively, then

$$M_w = \sigma_w Z_w = \sigma_w = bd^2/6$$
$$\text{And } M_s = \sigma_s Z_s = \sigma_s(td^2/6 \times 2)$$
$$\text{But } \sigma_s = \text{m}\,\sigma_w$$
$$\text{Therefore, } M_s = \text{m}\,\sigma_w \times td^2/3$$

The moment of resistance (M_r) of the composite beam is:

$$M_r = M_w + M_s$$
$$= \sigma_w \times \left[(\text{b} + 2\text{mt})\text{d}^2/6\right]$$

It can therefore be said that the moment of resistance of the flitched beam is the same as that of a wooden beam of width (b +2mt) and depth d. This beam is called an equivalent beam and its depth is the same as that of the given beam, while the width is increased by m times the thickness of the reinforcing steel plates.

Other types of flitch beams can be constructed with a steel plate between two wood beams or with two angles fitted around the edge of the wood beam as shown in Fig. 8.11c and d. It should be noted, however, that due to the rising cost of labour, the use of this type of beam has greatly declined. The advent of high-strength engineered lumber, has rendered this system largely obsolete.

8.5.3 Lumber Decks

Lumber is commonly used in the construction of floor, roof and bridge decks. The decking material may be solid-sawn wood, nailed-laminated material or glulam. Solid-sawn wood decking (Fig. 8.12) is typically fabricated with nominal 50-, 75- and 100-mm-thick members, while the widths are usually nominal 150 through 300 mm. In the case of nailed-laminated decking, planks often rough sawn and unseasoned are placed on edge on supports and nailed or bolted together horizontally. When situations call for deck thicknesses over nominal 50 mm, glulam with 3–5 laminations may be used.

The design of decking for stress and deflection is usually based on the assumption of uniformly distributed loads and equally spaced supports. The applicable design formulas for the common types of deck configurations are presented in Table 8.3. These formulas are used for decks that are flat or with less than 3 in 12 pitch, or with moderate camber to permit proper drainage in the case of roof

Fig. 8.12 A lumber deck

Table 8.3 Design formulas for lumber decking

Deck type	Bending stress (N/mm^2)	Deflection (mm)
Simple span	$\frac{wL^2c}{8I}$	$\frac{5wL^4}{384EI}$
Two-span continuous	$\frac{wL^2c}{8I}$	$\frac{wL^4}{185EI}$
Combined simple and two-span	$\frac{wL^2c}{8I}$	$\frac{wL^4}{109EI}$
Cantilevered pieces Inter-mixed	$\frac{wL^2c}{6.66I}$	$\frac{wL^4}{105EI}$

Note L span (mm), c half of deck thickness (mm), I moment of inertial of a 300-mm wide section of deck (mm^4), w maximum load N/mm length of deck

decks. In these cases, 'w' is the load per millimetre for a 300-mm (one-foot)-wide section, while 'I' is the moment of inertia for a 300-mm wide section.

8.5.4 Stressed-Skin Panels

A stressed-skin panel is a structural component consisting of longitudinal members fabricated with lumber stringers to which is bonded an OSB or a plywood cover (skin) either on one or both faces of the panel (Fig. 8.13a, b). Though the skins are most often made from plywood and OSB, other building materials such as gypsum wall board, sheetrock, wafer board and sheet metal are used as well. The exterior skin must be a nailable material since the panels are attached to the building's exterior wall and roof panels. They are also used as floor panels and foundation walls. Panels are usually manufactured in factories under controlled conditions to standardize quality. Panels with top and bottom skins are most common, while one-sided panels are generally used in special situations. A variation of the single panel with lumber strips on the bottom of the stringers (Fig. 8.13c) is called a T-flange panel.

The skin, if adequately bonded to the lumber stringers will, in combination with them, forms a T-beam with a considerably greater moment of inertia and section modulus than possessed by the stringers alone. The panel skin(s) take most of the bending stresses and perform a sheathing function, while the lumber stringers take the shear stresses. For maximum stiffness, a rigid connection (preferably provide by gluing) must be provided between the skin and the lumber stingers. Additional elements may be included in the panel such as headers at the end of the panel and blocking within the panel, which serve to align the stringers, support skin edges and help to distribute concentrated loads, as well as transverse stiffeners which help to stiffen the plywood skin under transverse bending and to resist transverse buckling when the skin is in longitudinal compression due to longitudinal bending loads.

The design loads for stressed-skin panels must not be less than required by the governing building regulations. Allowance must also be made for any temporary erection loads or moving concentrated loads. For design purposes, a trial section must first be assumed and then checked for its ability to sustain the expected loads.

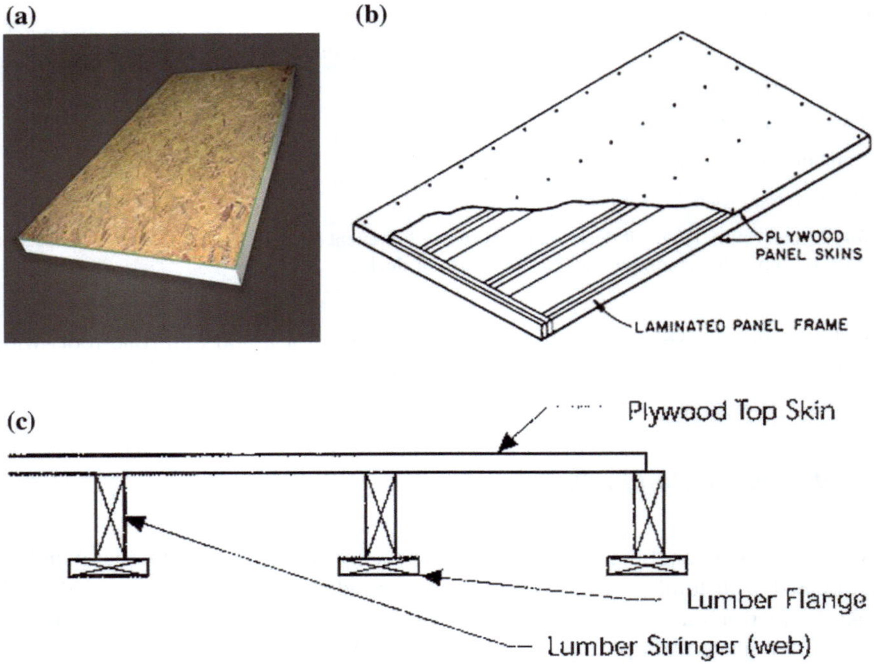

Fig. 8.13 **a** A two-sided OSB-overlaid stressed-skin panel. **b** Schematic diagram of a two-sided plywood-overlaid stressed-skin panel. **c** Front view of a typical T-flange stressed-skin panel

The design is usually controlled by deflection, since the panels are relatively shallow. Hence, adequacy in deflection should be checked first, followed by bending moment and then shear stress which is the least likely design parameter to control stressed-skin panel design. Shear may, however, control the design if one or both skins are thick and the span is short.

Also, shear deflection should be considered separately for stressed-skin panels and added to bending deflection, unlike in most timber design situations where it is automatically taken into account in the modulus of elasticity. In addition to computing the deflection of the whole panel, acting as a unit, the deflection of the top skin between the stringers must also be checked. The basic formulas and the computation procedures are available in design codes for stressed-skin panels.

Worked Examples on Wooden Beams

1. It is proposed to design a wooden platform using wooden floor joists. The span of each of the floor joists is 3500 mm, and it is to be assumed as being simply supported at both ends. The joists are to be of rectangular section. Given that the expected live load on the platform is equivalent to 0.5 N/mm of span, uniformly distributed, and the design values stated below, design the joist.

Design Values

Bending strength of the wood = 11.20 N/mm^2
Shear strength parallel to the grain of the wood, F_v = 1.40 N/mm^2
Compression perpendicular to the grain of the wood = 2.80 N/mm^2
Modulus of elasticity of the wood species, E = 9500 N/mm^2
Allowable deflection = $L/360$ (mm) for floor joists carrying live load only.

Solution Steps

(i) **Sketch the beam based on the information available/given.**

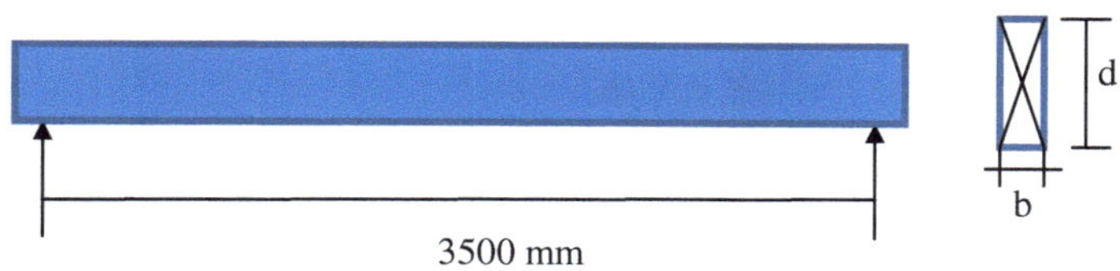

(ii) **State the available design parameters and tabulated values**

Span of each joist = 3500 mm
Expected live load on the platform = 0.5 N/mm
For this particular example, let us assume that all modification factors are unity (i.e. 1).

(i) **Select a trial cross-section**, i.e.
Assuming a rectangular cross-section, b = 75 mm, d = 100 mm.

(ii) **Compute the following design parameters**:

(a) **The maximum bending moment (M)**

$$M = \frac{wL^2}{8} \quad \text{for uniformly distributed load}(w)\text{ on a simply supported beam}$$
$$= 0.5 \times (3500)^2/8 = 765625 \text{ Nmm}$$

(b) **The Section Modulus (S)**
This can be calculated as follows or read from Table 5.1 in Chap. 5.
For a rectangular cross-section of depth 'd' and width 'b':

$$S = bd^2/6$$
$$= 75 \times (100)^2/6 = 1,25,000 \text{ mm}^3$$

(c) **Moment of Inertial (edgewise use is assumed)**
This can be calculated or read from Table 5.1 in Chap. 5.
The tabulated value, $I = 6{,}250{,}000\ \text{mm}^4$

(d) **The maximum shear force (V) and shear stress (f_V) on the beam**
For a simply supported, single span beam of span L, sustaining a uniformly distributed load (w), the maximum shear force,

$$V = \frac{w \cdot L}{2}$$

$$V = 0.5 \times 3500 \times \frac{1}{2} = 875\ \text{N}$$

$$\begin{aligned} f_V &= 1.5 \times \text{V}/A \\ &= 1.5 \times \text{V}/bd \\ &= 1.5 \times 875/(75 \times 100) = 0.175\ \text{N/mm}^2 \end{aligned}$$

(e) **Allowable deflection (Δ_{all})**

$$\begin{aligned} \Delta_{all} &= L/360(\text{mm})\ \text{for floor joists carrying live load only} \\ &= 3500/360 = 9.72\ \text{mm} \end{aligned}$$

(f) **Expected deflection (Δ_{max})**
For a simply supported, single span beam of span L, under uniform loading (w), the maximum deflection is:

$$\begin{aligned} \Delta_{max} &= \frac{5wL^4}{384EI} \\ &= \frac{5 \times 0.5 \times (3500)^4}{384 \times 9500 \times 6{,}250{,}000} \\ &= 16.45\ \text{mm} \end{aligned}$$

- **NOTE**: Since $\Delta_{max} > \Delta_{all}$, we have to select a greater cross-section, e.g. 75 mm × 150 mm, and re-compute Δ_{max}
For a 75 mm × 150 mm, $I = 21{,}093{,}750.0\ \text{mm}^4$

$$\Delta_{max} = \frac{5 \times 0.5 \times (3500)^4}{384 \times 9500 \times 21093750.0}$$

$$\boldsymbol{\Delta_{max} = 4.88\ \text{mm}}$$

- This is now OK.
- **From now onwards, we assume a cross-section of 75 mm × 150 mm.**

Use the beam design governing equations, applying the appropriate modification factors, to check for compliance with bending stress, shear stress, deflection and bearing stress criteria.

(a) **Bending Stress Criterion**

Bending Strength of the wood ≥ Induced Bending Stress in the beam, i.e.,

$$\sigma_b \geq M/S$$

Tabulated Bending Strength of the timber grade selected, σ_b = 11.20 N/mm^2

$$M = 7,65,625 \text{ Nmm}$$
$$\textbf{New S} = 281,250 \text{ mm}^3$$
$$M/S = 765,625/281,250 = 2.72 \text{ N/mm}^2$$

- **Since σ_b > *M/S*, the joist is OK in Bending**

(b) **Shear Stress Criterion**

Shear strength of the wood ≥ Induced Shear stress in the beam, i.e.,

$$F_v \geq f_v$$

$$\textbf{New}\, f_V = 1.5 \times V/A$$
$$= 1.5 \times 875/75 \times 150 = 0.12 \text{ N/mm}^2$$

Tabulated shear strength of the timber grade selected, F_v = 1.40 N/mm^2

$$f_v = 0.12 \text{ N/mm}^2$$

- **Since $F_v \geq f_v$, the joist is OK in shear**

(c) **Deflection Criterion**

Allowable deflection (deflection limit) ≥ Maximum induced deflection, i.e.

$$\Delta_{all} \geq \Delta_{max}$$
$$\Delta_{all} = 9.72 \text{ mm}$$
$$\Delta_{max} = 4.88 \text{ mm}$$

- **Since $\Delta_{all} \geq \Delta_{max}$, the joist is OK in deflection**

Design Conclusion

A set of 75 mm × 150 mm joists will carry the expected uniformly distributed live load satisfactorily.

(d) **Check for lateral stability**

1. The ratio of depth-to-breadth (*d*/*b*) based on nominal dimension is 150/75, i.e. 2 to 1. Therefore, no lateral support shall be required.
2. Use Microsoft Excel to carry out the design of a typical floor joist for the same wooden platform.

Solution

1. Open an Excel worksheet.
2. Input the given design information into an Excel worksheet as shown in Fig. 8.14a
3. Select a trial section, $b \times d$ and input the values.
4. Use Excel to compute the design parameters, using cell references.
5. The result is as presented in Fig. 8.14b.

3. Develop a Microsoft Excel template for the use in the design of wooden beams. The template should have drop-down menu buttons for selecting pre-stored design values.

Solution

(i) Open an Excel worksheet 1.
(ii) Open another Excel worksheet 2. Input the given design information obtained from a structural wood design code into as shown in Fig. 8.15a. This is referred to as creating a drop-down list.
(iii) To add this drop-down list to the worksheet 1 already opened, do the following:

- Choose validation from the data menu as shown in Fig. 8.15b.
- Choose list from the allow option's drop-down list as shown in Fig. 8.15c.
- Click the Source control and drag to highlight the cells A1:A4. …
- Make sure the n-cell drop-down option is checked. …
- Click OK.

(iv) The wooden beam design template obtained is as presented in Fig. 8.11d.

4. A flitch beam consists of a wooden joist 100 mm wide, 200 mm deep strengthened by two steel plates 80 mm thick and 200 mm deep, one on either side of the joist. If the stresses in wood and steel are not to exceed 7.5 and 125 N/mm^2, find the moment resistance of the beam. Assume that the ratio of the modulus of elasticity of steel to that of wood is 20.

(a)

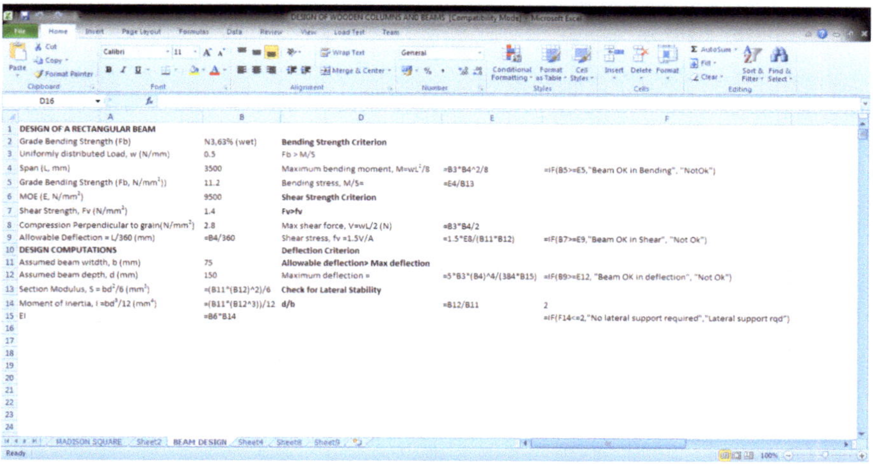

(b)

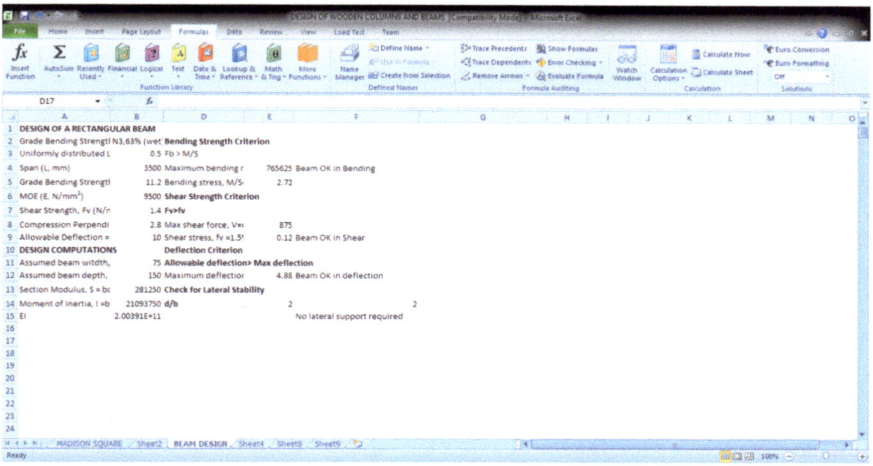

Fig. 8.14 **a** Excel worksheet showing the design computations. **b** Excel worksheet showing the results

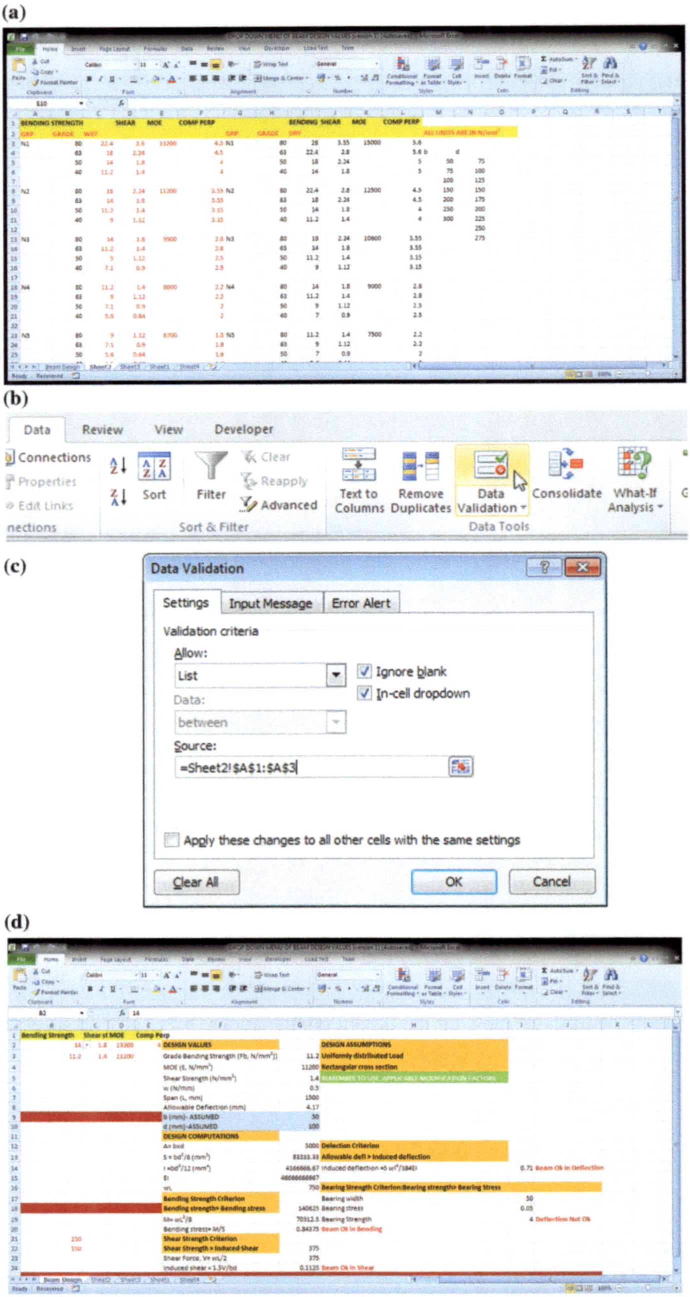

Fig. 8.15 **a** Creating a drop-down list of design values. **b** Selecting validation from the data menu. **c** Selecting 'list' from the allow option's drop-down list. **d** An excel wooden beam design template

Solution

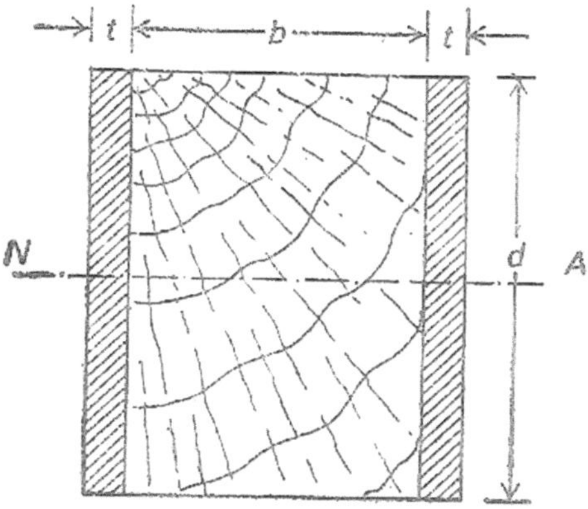

Width of the wooden joist, b = 100 mm
Thickness of the steel plates, t = 80 mm
Depth of the beam, d = 200 mm
Allowable stress in wood, σ_w = 7.5 N/mm^2
Allowable stress in steel plates, σ_s = 125 N/mm^2
Ratio of the modulus of elasticity of steel to wood, m = 20
Moment resistance of the beam, M_r = Resistance of wood (M_w) + Resistance of steel (M_s).

$$= \sigma_w \times \left[(b + 2\text{mt})d^2/6\right]$$
$$= 7.5 \times [(100 + 2 \times 20 \times 80)]/6$$
$$= 165 \times 10^6 \text{N mm} = 165 \text{ Nm}$$

5. Write a simple FORTRAN program for selecting adequate span of wooden beams.

Solution

```
! Fortran program to select adequate span of rectangular
and square wooden beams
real::condition
write(*,*) "
This program can be used to select the adequate length of single span,
simply supported and cantilever wooden beams"
write(*,*) "subjected to either concentrated load at mid-span or
uniformly distributed
loads"
```

```
write(*,*) "concentrated load at mid-span on simply supported,
condition = 1"
write(*,*) "uniformly distributed load on simply supported beam,
condition = 2"
write(*,*) "concentrated load at extreme end on cantilevered beam,
condition = 3"
write(*,*) "uniformly distributed load on cantilevered  beam,
condition = 4"
write(*,*) "uniformly distributed load on can-
tilevered beam fixed at both ends,
condition = 5"
write(*,*) 'condition='
read(*,*)condition
write(*,*)'input the bending strength of the lumber in newton per square mm'
read(*,*)p
write(*,*)'input the shear strength of the lumber in newton per square mm'
read(*,*)q
write(*,*)'input  the modulus of elasticity of the lumber in new-
ton per square mm'
read(*,*)r
write(*,*)'input the length (span)of the beam in mm'
read(*,*)t
write(*,*)'input  the thickness/depth on the beam in mm'
read(*,*)v
write(*,*)'input  the width of the beam in mm'
read(*,*)b
write(*,*)'input  the deflection limit as a frac-
tion of the span of the beam in mm'
read(*,*)w
if (condition<=1) goto 5
 if (condition<=2)  goto 200
   if (condition<=3)  goto 600
     if (condition<=4)  goto 800
       if (condition<=5)  goto 910
5 write(*,*)'input a trial concentrated load on the beam in newton'
read(*,*)u
!c computations for wooden beams subjected to concentrated load
at mid-span
    bstress=(u*t)/4
   bstrenth =p*b*(v**2)/6
   det=bstrenth-bstress
   if (det)20,30,40
     20 go to 110
     30 go to 50
```

```
  40 go to 50

 50 sstress= (1.5*u*0.5)/(b*v)
     det2=q-sstress
    if (det2)80,90,100
    80 go to 125
    90  go to 105
    100  go to 105
  105 mmt=b*(v**3)/12
  defl=(u*(t**3))/(48*r*mmt)
  det3=w-defl
   if  (det3)120,130,140
   120 goto  145
   130 go to 150
   140 go to 150
    110 write(*,*)'selected load not acceptable,beam will fail
in bending, choose
another load/strength grp/grade'
      125 write(*,*)'selected load not acceptable,beam will fail
in shear, choose
another load/strength grp/grade'
      145 write(*,*)'selected load not acceptable,beam will fail
in deflection, choose
another load/strength grp/grade'
      stop
  150 write(*,*)'selected load is acceptable'
  write(*,*)'load (n) =',u
  write(*,*)'width of the beam =',b
  write(*,*)'depth of the beam =',v
  write(*,*)'span of the beam =',t
  write(*,*)'moment capacity of the beam =',bstrenth
              write(*,*)'induced moment on the beam =',bstress
                     write(*,*)'shear strength of the beam =',q
                     write(*,*)'induced shear stress =',sstress
                    write(*,*)'max allowable deflection =',w
                    write(*,*)'expected deflection =',defl
          write(*,*)'remember to check for beam stability'
          !c latsta=int(v/b)
          !c goto 1004
          stop
          goto 1000
              !c  computations for wooden beams subjected
to uniformly distributed
load
```

```
200 write(*,*)'input a trial uniformly distributed load on the beam
in newton per mm'
read(*,*)ul
   bstress=(ul*(t**2))/8
  bstrenth =p*b*(v**2)/6
  det=bstrenth-bstress
  if (det)220,230,240
    220 go to 310
    230 go to 250
    240 go to 250

  250 sstress= (0.5*t*ul)/(b*v)
      det2=q-sstress
     if (det2)280,290,300
    280 go to 325
     290  go to 305
     300  go to 305
   305 mmt=b*(v**3)/12
   defl=(5*ul*(t**4))/(384*r*mmt)
   det3=w-defl
   if  (det3) 360,370,380
     360 goto  400
     370 go to 500
     380 go to 500
      310 write(*,*)'selected load not acceptable,beam will fail
in bending, choose
another load/strength grp/grade'
      go to 1000
        325 write(*,*)'selected load not acceptable,beam will fail
in shear, choose
another load/strength grp/grade'
         go to 1000
        400 write(*,*)'selected load not acceptable,beam will fail
in deflection,choose
another load/strength grp/grade'
         go to 1000
                 500 write(*,*)'selected load is acceptable'
                   write(*,*)'load (n/mm) =',ul
                  write(*,*)'span of the beam =',t
                 write(*,*)'width of the beam =',b
                 write(*,*)'depth of the beam =',v
                 write(*,*)'moment capacity of the beam =',bstrenth
             write(*,*)'induced moment on the beam =',bstress
                       write(*,*)'shear strength of the beam =',q
                       write(*,*)'induced shear stress =',sstress
```

```
                  write(*,*)'max allowable deflection =',w
                  write(*,*)'expected deflection =',defl
                write(*,*)'remember to check for beam stability'
        !c latsta=int(v/b)
        !c goto 1004
        stop
        goto 1000
  !c  computations for cantilever beam subjected to point load at one end
 600 write(*,*)'input a trial point load on the cantilever beam in newton'
read(*,*)wl
!c computations
   bstress=wl*t
  bstrenth =p*b*(v**2)/6
  det=bstrenth-bstress
  if (det)605,610,615
   615 go to 650
   605 go to 750
   610 go to 650

  650 sstress= (1.5*wl)/(b*v)
     det2=q-sstress
     if (det2)665,670,675
     665 go to 780
     670  go to 680
     675  go to 680
   680 mmt=b*(v**3)/12
   defl=(wl*(t**3))/(3*r*mmt)
   det3=w-defl
    if  (det3)685,690,695
    685 goto  790
    690 go to 795
    695 go to 795
    750 write(*,*)'selected load not acceptable,beam will fail
in bending, choose
another load/strength grp/grade'
    goto 1000
      780 write(*,*)'selected load not acceptable,beam will fail
in shear, choose
another load/strength grp/grade'
      790 write(*,*)'selected load not acceptable,beam will fail
in deflection, choose
another load/strength grp/grade'
      goto 1000
      795 write(*,*)'selected load is acceptable'
```

```
      write(*,*)'load (n/mm) =',wl
              write(*,*)'span of the beam =',t
         write(*,*)'width of the beam =',b
          write(*,*)'depth of the beam =',v
      write(*,*)'moment capacity of the beam =',bstrenth
            write(*,*)'induced moment on the beam =',bstress
                    write(*,*)'shear strength of the beam =',q
                    write(*,*)'induced shear stress =',sstress
                   write(*,*)'max allowable deflection =',w
                   write(*,*)'expected deflection =',defl
         write(*,*)'remember to check for beam stability'
          !c latsta=int(v/b)
          !c goto 1004
          stop
          goto 1000
    !c  computations for cantilever beam subjected to uniformly
distributed load
800 write(*,*)'input a trial uniformly distributed load on the
cantilever beam in
newton per mm'
read(*,*)udl
!c computations
  bstress=udl*(t**2)*0.5
  bstrenth =p*b*(v**2)/6
  det=bstrenth-bstress
  if (det)805,810,815
   815 go to 835
   805 go to 900
   810 go to 835


  835 sstress= (1.5*udl*t)/(b*v)
     det2=q-sstress
     if (det2)840,845,850
     840 go to 880
     845  go to 860
     850  go to 860
   860 mmt=b*(v**3)/12
   defl=(udl*(t**4))/(8*r*mmt)
   det3=w-defl
    if  (det3)865,870,875
     865 goto  890
     870 go to 895
     875 go to 895
     900 write(*,*)'selected load not acceptable,beam will fail
```

```
in bending, choose
another load/strength grp/grade'
    goto 1000
       880 write(*,*)'selected load not acceptable,beam will fail
in shear, choose
another load/strength grp/grade'
       goto 1000
       890 write(*,*)'selected load not acceptable,beam will fail
in deflection, choose
another load/strength grp/grade'
       goto 1000
       895 write(*,*)'load is acceptable'
       write(*,*)'load (n/mm) =',udl
       write(*,*)'span of the beam =',t
          write(*,*)'width of the beam =',b
          write(*,*)'depth of the beam =',v
          write(*,*)'moment capacity of the beam =',bstrenth
              write(*,*)'induced moment on the beam =',bstress
                       write(*,*)'shear strength of the beam =',q
                       write(*,*)'induced shear stress =',sstress
                      write(*,*)'max allowable deflection =',w
                      write(*,*)'expected deflection =',defl
          write(*,*)'remember to check for beam stability'
          !c latsta=int(v/b)
          stop
          goto 1000
          !c goto 1004
            !c  computations for cantilever beam fixed at both endsm
subjected to udl
         910 write(*,*)'input uniformly distributed load on the
cantilever beam in
newton per mm'
read(*,*)udlc
!c computations
  bstress=udlc*(t**2)/12
  bstrenth =p*b*(v**2)/6
  det=bstrenth-bstress
  if (det)915,920,925
   915 go to 990
   920 go to 930
   925 go to 930
```

```
  930 sstress= (1.5*udlc*t*0.5)/(b*v)
      det2=q-sstress
     if (det2)935,940,945
     935 go to 980
     940  go to 950
     945  go to 950
   950 mmt=b*(v**3)/12
   defl=(udlc*(t**4))/(384*r*mmt)
   det3=w-defl
    if  (det3)960,965,970
     960 goto  999
     965 go to 975
     970 go to 975
     990 write(*,*)'selected load not acceptable,beam will fail
in bending, choose
another load/strength grp/grade'
     goto 1000
         980 write(*,*)'selected load not acceptable,beam will fail
in shear, choose
another load/strength grp/grade'
         goto 1000
         999 write(*,*)'selected load not acceptable,beam will fail
in deflection, choose
another load/strength grp/grade'
         goto 1000
         975 write(*,*)'load is acceptable'
              write(*,*)'load (n/mm) =',udlc
              write(*,*)'span of the beam =',t
               write(*,*)'width of the beam =',b
               write(*,*)'depth of the beam =',v
               write(*,*)'moment capacity of the beam =',bstrenth
                     write(*,*)'induced moment on the beam =',bstress
                                   write(*,*)'shear strength of the beam =',q
                                   write(*,*)'induced shear stress =',sstress
                                  write(*,*)'max allowable deflection =',w
                                  write(*,*)'expected deflection =',defl
               write(*,*)'remember to check for beam stability'
               !c latsta=int(v/b)
             !c  1004 if (int(latsta)<=2) goto 1005
             stop
               goto 1000
 1000 end
```

Practice Questions

1. Which assumption regarding the bending equation of a beam is seldom satisfied in practice? Why is it not satisfied often?
2. Comment on the typical deflection behaviour of a wooden beam under long duration load.
3. Why is it necessary to control the deflection of a wooden beam?
4. A wooden cantilever beam, 100-mm wide projects 1.5 m where it is loaded with pulley tackle and load (including the downward pull of the operator) of 2 kN. Its bending strength may be 6.6 N/mm^2, find a suitable depth.
5. Find a suitable depth for a domestic floor joist 50 mm thick if the permitted bending stress is 8.6/mm^2, clear span = 4.5 m, spacing = 400 mm and loading = 2 kN/m^2.
6. A 35-m-long wooden beam made from *Terminalia superba* is to carry a load of 5 kN at mid-span at an angle of 30◦ to the horizontal. If the beam is wet and the grade is 50%, select an appropriate size (dimension) for the beam. Assume the following design values:

   Bending strength of the wood = 4.50 N/mm^2
   Shear strength parallel to the grain of the wood, F_v = 0.49 N/mm^2
   Compression perpendicular to the grain of the wood = 1.25 N/mm^2
   Modulus of Elasticity of the wood species, E = 5600 N/mm^2
   Allowable deflection = $L/360$ (mm).

7. Determine if a 6-m-long 75 mm × 100 mm wooden beam is adequate for a dead load of 5 kN, if the beam is to be fabricated from a timber in dry condition. Assume a load duration factor of 0.9, an allowable deflection limit of $L/180$ and the following design values:

   Bending strength of the wood = 28. 0 N/mm^2
   Shear strength parallel to the grain of the wood, F_v = 3.55 N/mm^2
   Compression perpendicular to the grain of the wood = 5.60 N/mm^2.
   Modulus of elasticity of the wood species, E = 9500 N/mm^2.

8. Design a roof beam 4-m long to support a uniformly distributed dead load of 3.5 kN/m^2. Beams are spaced 400 mm on centre and sufficient roof slope is provided to prevent ponding. The wood material is *Terminalia superba*, 80% grade. Assume dry service condition and the following design values:

   Bending strength of the wood = 9.0 N/mm^2
   Shear strength parallel to the grain of the wood, F_v = 1.12 N/mm^2
   Compression perpendicular to the grain of the wood = 1.80 N/mm^2
   Modulus of Elasticity of the wood species, E = 6300 N/mm^2
   Allowable deflection = $L/240$ (mm)

(Hint: uniform loads are obtained by multiplying the given design loads by the tributary width, e.g., dead load = 3.5 kN/m^2 × 0.4 m = 1400 N/m).

9. A simply supported 6-m-long wooden beam is expected to carry a concentrated load of 2000 N at mid-span. Determine the size of lumber required assuming a load duration factor of 0.9 and wet service condition. Take the maximum deflection as $L/360$ and assumed that *Melicia excelsa* of 50% grade (wet) will be used. Assume the following design values:

   Bending strength of the wood = 9.0 N/mm^2
   Shear strength parallel to the grain of the wood, F_v = 1.12 N/mm^2
   Compression perpendicular to the grain of the wood = 2.50 N/mm^2
   Modulus of Elasticity of the wood species, E = 9500 N/mm^2.

10. A 3.6-m-long log of wood is to be used as a bridge across a stream. If the estimated uniformly distributed load is 1.5 N/mm, select the appropriate diameter. Assume the following design values:

    Bending strength of the wood = 7.10 N/mm^2
    Shear strength parallel to the grain of the wood, Fv = 0.9 N/mm^2
    Compression perpendicular to the grain of the wood = 1.8 N/mm^2
    Modulus of Elasticity of the wood species, E = 6700 N/mm^2.

11. Determine the allowable load/mm for a 5-m-long, 75 mm × 125 mm wooden beam, assuming a 40% grade timber in dry condition with the following design values is used.

    Bending strength of the wood = 14.0 N/mm^2
    Shear strength parallel to the grain of the wood, F_v = 1.8 N/mm^2
    Compression perpendicular to the grain of the wood = 5.0 N/mm^2
    Modulus of Elasticity of the wood species, E = 15,000 N/mm^2.

12. A wet *Afzelia Africana* (50% grade) beam has a cross-section of 100 mm 200 mm. The beam is to carry a uniformly distributed load of 3.5 N/mm. Determine the allowable span of the beam. Assume the following design values:

    Bending strength of the wood = 11.2 N/mm^2
    Shear strength parallel to the grain of the wood, F_v = 1.4 N/mm^2
    Compression perpendicular to the grain of the wood = 3.15 N/mm^2
    Modulus of elasticity of the wood species, E = 11,200 N/mm^2.

13. A simply supported wooden beam is 3 m long and carries a uniformly distributed load of 150 N/m of length. Determine the cross-section of the beam if:

    (a) It is to be square.
    (b) The depth is to be 1-1/2 times the width.
    (c) The depth of the beam is to be 1-1/4 the width.

Assume the following design values:
Bending strength of the wood = 9.0 N/mm^2
Shear strength parallel to the grain of the wood, F_v = 1.12 N/mm^2
Compression perpendicular to the grain of the wood = 2.50 N/mm^2
Modulus of elasticity of the wood species, E = 9500 N/mm^2.

14. The joists for a 7.2-m-span office floor system are to be designed and fabricated. Each joist is expected to carry a uniformly distributed live load of 0.75 N/mm and a dead load, including an allowance for the self-weight of the members of 0.15 N/mm.

    (a) Assuming simple supports at both ends, and a load duration factor of 0.9, design a typical joist based on the following design values:

    Bending strength of the wood = 14.0 N/mm^2
    Shear strength parallel to the grain of the wood, F_v = 1.8 N/mm^2
    Compression perpendicular to the grain of the wood = 5.0 N/mm^2
    Modulus of Elasticity of the wood species, E = 15,000 N/mm^2.

    (b) If the width of the office is 4.8 m and joists are to be spaced 240 mm on centre, compute the number of joists required.

15. Develop an Excel template for the design of square and circular wooden beams in different loading conditions.

References

Breyer DE, Fridley KJ, Cobeen KE (1999) Design of wood structures, 4th edn. McGraw-Hill, New York, USA

Faherty KF, Williamson TG (1999) Wood engineering and construction handbook, 3rd edn. McGraw-Hill Handbooks, New York, USA

Goldstein EW (1999) Timber construction for architects and builders. McGraw-Hill Publishers, New York, USA

Gurfinkel G (1973) Wood engineering. Southern Forest Products Association, USA

Khurmi RS, Khurmi N (2013) Strength of materials (mechanics of solids in S.I. units, international edition). S. Chand & Company PVT Ltd., New Delhi

Lindley JA, Whitaker JH (1996) Agricultural buildings and structures, revised edition. The American Society of Agricultural Engineers, USA

Marshall WT, Nelson HM (1978) Structures. Pitman Publishing Ltd., London

Midwest Plan Service (1977) Structures and environment handbook, published b Midwest Plan Service. Ames, Iowa

NCP (2005) Nigerian code of practice for structural timber design. The Standards Organisation of Nigeria

Ozelton EC, Baird JA (1976) Timber designers' manual. Granada Publishing Limited, London

Seward D (2003) Understanding structures, analysis, materials, design, 3rd edn. Palgrave Macmillan, New York, USA

Shaeffer RE (1999) Elementary structures for architects and builders, 4th edn. Prentice-Hall Publishing Co., New Jersey, USA

Singh S (1982) Strength of materials, 2nd revised ed. Vikas Publishing House PVT Ltd., U.P.
Willenbrock JH, Manbeck HB, Suchar MG (1998) Residential building design and construction. Prentice Hall, New Jersey
Wolfe R (1998) Structural use of small-diameter round timbers, USDA Forest Service, Forest Products Laboratory (FPL) publication No. 72, Vol. 3, Madison, Wisconsin

Chapter 9
Design of Solid Wood Columns

9.1 General Classification of Columns

Columns are essentially compression elements used to support loads that tend to shorten them. Examples of columns include studs or poles holding a roof, chair, machine legs, ladder, side rails. Columns in general can be classified on the basis of construction material, shape and slenderness ratio. For example, based on the type of construction material, we have steel, concrete, wooden (solid or built-up) columns; based on shape, we have rectangular, square, circular, I-shaped, T-shaped, tapered and box columns; and on the basis of slenderness ratio we have short, intermediate and long columns.

Columns for buildings are usually designed with greater factors of safety than other structural elements because any column failure would result in the catastrophic collapse of at least a major portion of the building frame. When a column fails, the beams or girders framing into it come down, as do all the other columns directly above. The great majority of wooden columns are solid and rectangular in cross-section. They are designed as simple members with hinged ends, free to rotate but not free to move laterally. When sufficiently overloaded, columns fail either by ***buckling*** (Fig. 9.1a) or ***crushing*** (Fig. 9.1b) or a combination of these two effects.

Regardless of the material of construction, very short, stout columns tend to fail by crushing, while long, slender columns tend to fail by buckling. However, most columns in buildings are proportioned such that both effects would be involved. It must be re-emphasised that the relative susceptibility of columns to buckling is a function of their slenderness ratio. In its traditional form, slenderness ratio is expressed as the effective unbraced length of a column divided by the least radius of gyration, i.e. L_e/r_{min} (we shall talk more about effective length shortly). However, the slenderness ratio of a rectangular column, the most common type of wooden columns, is usually taken as the ratio of effective unbraced length to the least lateral dimension (thickness), i.e. slenderness ratio = L_e/d_{min}.

A.O. Olorunnisola, *Design of Structural Elements with Tropical Hardwoods*,
DOI 10.1007/978-3-319-65343-3_9

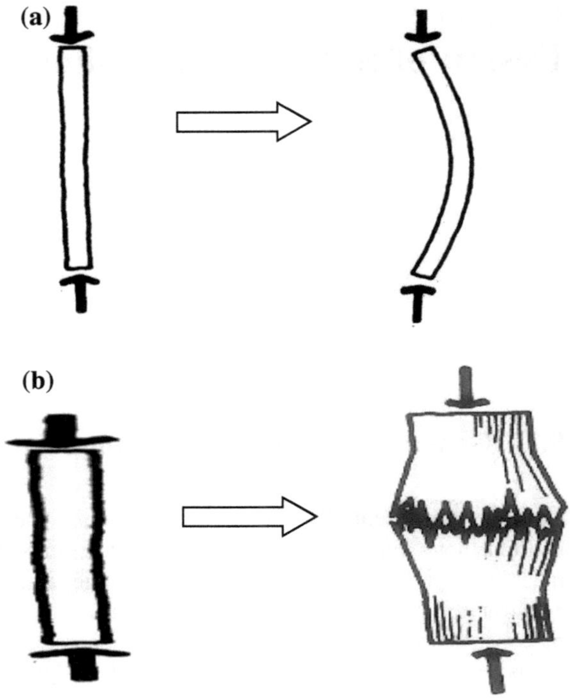

Fig. 9.1 **a** Buckling of a column. **b** Crushing of a column

Short wooden columns generally have relatively small slenderness ratios, i.e.

$$\mathbf{L_e/r_{min} \leq 38 \quad or \quad L_e/\,d_{min} \leq 11}$$

They tend to fail by crushing parallel to the grain. Examples of short columns are shown in Fig. 9.2.

Intermediate wooden columns (Fig. 9.3) tend to have slenderness ratio (L/r_{min}) between 38 and k_r, or slenderness ratio (L_e/d_{min}) between 11 and k.

Both k and k_r are numbers dependent on the elastic modulus and the compression strength of the column material, i.e.

$$38 < \frac{L_e}{r_{min}} \leq k_r,\ \text{where}\ k_r = 2.32\sqrt{E/\sigma_G},\ \text{or}$$

$$11 < \frac{L_e}{d_{min}} \leq k,\ \text{where}\ k = 0.671\sqrt{E/\sigma_G}$$

Fig. 9.2 Wooden posts used as short columns in fencing

Fig. 9.3 Lumber used as intermediate columns

where

E = modulus of elasticity (N/mm^2) and
σ_G = compression strength parallel to the grain of the column material (N/mm^2).

Long wooden columns (Fig. 9.4) tend to have slenderness ratios exceeding k or k_r 9r subscript), i.e. $L_e/r_{min} \geq k_r$ or $L_e/d_{min} \geq k$.

Long wooden columns will buckle at loads which are too low to cause compressive crushing.

Fig. 9.4 Wooden poles used as long columns in a building

9.2 End Conditions in Columns

How the column ends are connected to the rest of the structure has a large influence on the initial buckling load. Typically, column end conditions are identified in various fields of structural design. The most common ones are those with:

- Both ends pinned as shown in Fig. 9.5.
- One end pinned the other fixed as shown in Figs. 9.6 and 9.7.
- Both ends fixed.
- One end fixed, the other free as typified by an electric pole.

Fixed end condition means that the joint connection is so stiff that no member meeting at the joint can rotate independently of the other member(s). In the case of a pinned end condition, members meeting at the joint can rotate independently of one another. In many instances, the possibility of side sway in also considered. Side sway means that the top of the column is relatively free to displace laterally with respect to the bottom of the column. Hence, there are situations in which provision is made for side sway and vice versa. If the column ends are fixed and thereby restrained from rotation in some manner, the effective buckling length can be very different from the true length. Theoretically, there is no difference between the actual length and the effective unbraced length of a pinned end column. For columns with other end conditions, however, the effective unbraced length is taken as the distance between inflection points on a sketch of the buckled column as shown in Fig. 9.8a–d. An inflection point corresponds to a point of reverse curvature on the deflected shape of the column and represents a point of zero moment. In general, the effective unbraced length of a column is determined mathematically by multiplying the effective length factor, k (the values of which are usually specified in design codes), by the actual unbraced length, i.e. effective length, L_e = effective length factor (k) x unbraced length.

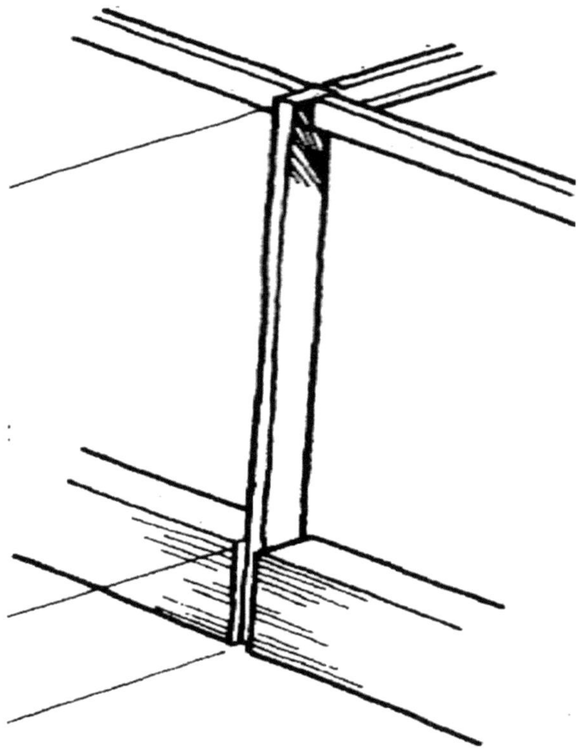

Fig. 9.5 Wooden column with pinned ends

Typically, six "ideal" column end conditions are identified in various fields of structural design, i.e.,

1. Both ends fixed, no side sway
2. One end pinned the other fixed, no side sway
3. Both ends pinned, no side sway
4. Both ends fixed, side sway allowed
5. One end fixed, the other free, side sway allowed
6. One end fixed the other pinned, side sway allowed

For end conditions in which sideway is prevented, the effective length of the column is less than or equal to its actual un-braced length (since $k \leq 1.0$). For end conditions in which side sway occurs, the effective length of the column is greater than its unbraced length (since $k > 1.0$). Generally, wooden columns are assumed to have 'square-cut ends' which offer some restraint against rotation at both ends. However, most practical column ends are not exactly square, and some accidental eccentricity may be present due to non-uniform bearing. These effects are, however, often assumed to be compensating. Therefore, a typical wooden column is assumed to be pinned at both ends and the effective length is assumed to

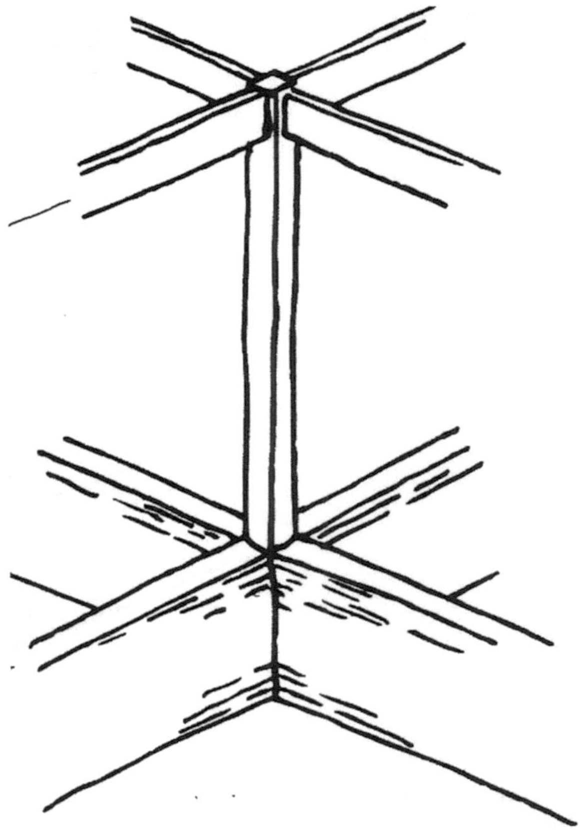

Fig. 9.6 Wooden column pinned at the *Top* and fixed at the *Bottom*

be equal to the unbraced length. It is important to also note that effective lengths are theoretical. Practical field column end conditions can only approximate the ideal pinned and fixed column end conditions. The designer must examine the actual bracing conditions and determine which 'ideal' column most closely approximates the actual end conditions for a given column. The designer must also be able to identify which columns will undergo sideway and what constitutes restraint against sideway. Some degree of judgment is required for this evaluation.

9.3 The Euler Column Buckling Equation

The basic theory of elastic buckling was successfully formulated by Leonhard Euler (1707–1783), a Swiss Mathematician who postulated that elastic buckling occurs because there is more than one position of equilibrium for a long, straight compression member, i.e., a slightly deflected column can still support a load and be in equilibrium just like the straight one. However, as the axial load is gradually

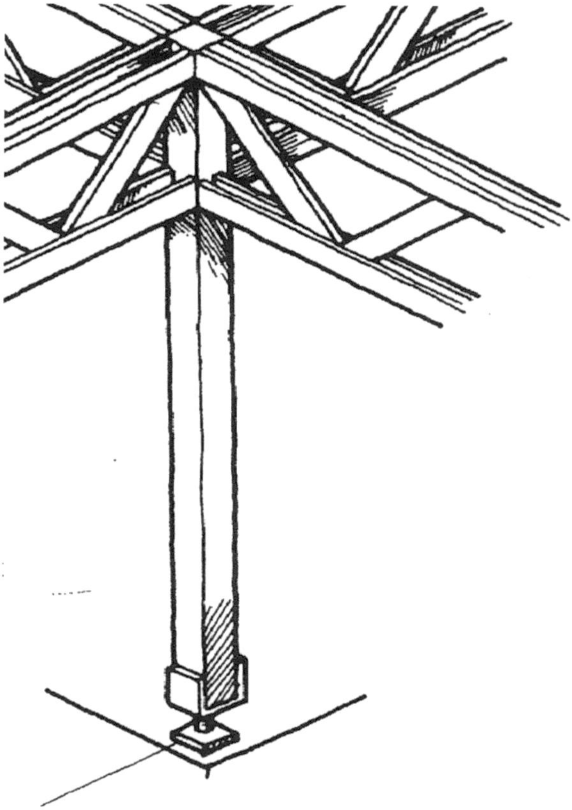

Fig. 9.7 Wooden column pinned at the *lower* end and fixed at the *upper* by deep trusses framing into it

increased on an initially straight long column, it will suddenly deflect laterally. If the load is removed, the column will return to its initial straight shape. The particular value of axial load (called the critical load) that causes buckling is given by the Euler equation:

$$P_{cr} = \frac{\pi^2 EI}{l^2}$$

where

P_{cr} = axial load necessary to cause buckling (N);
E = modulus of elasticity of the column material (Pa);
I = moment of inertia of the column cross-section (mm^4); and
l = length of the column (mm).

Just as in a beam, the rate of change of slope of a column is directly proportional to the bending moment and inversely proportional to the stiffness.

As shown in Fig. 9.9b, this relationship is given by the equation:

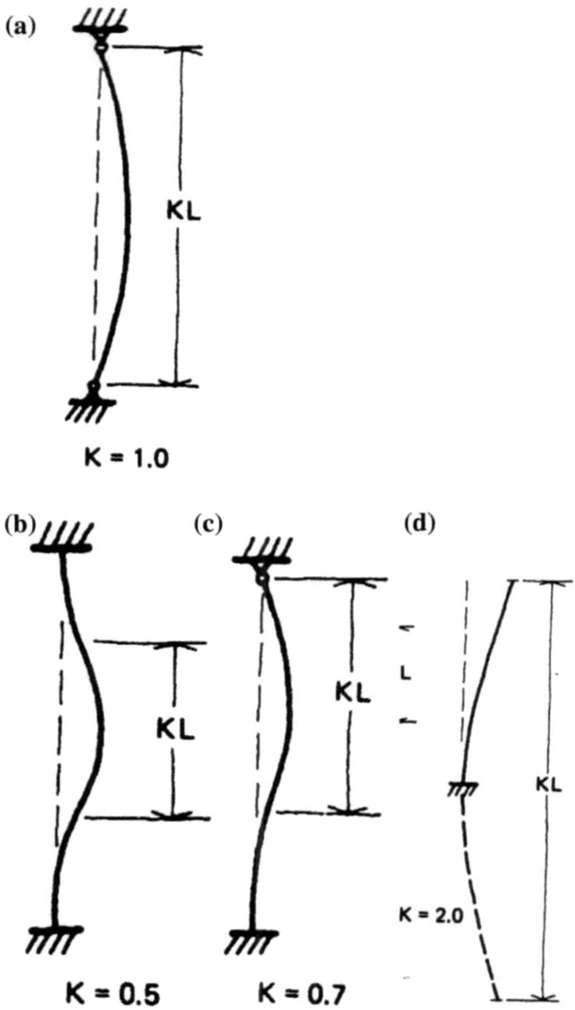

Fig. 9.8 Effective lengths for different column end conditions. *Note* a = pinned-pinned ends; b = fixed-fixed ends; c = fixed-pinned ends; d = fixed-free ends

$$\frac{-d^2y}{dx^2} = \frac{Py}{EI}$$

(The negative sign is present because of the selection of the origin for coordinate axes). As x increases, the slope decreases; thus, the rate of change of slope is negative. The equation can be rewritten as:

$$\frac{d^2y}{dx^2} + \frac{Py}{EI} = 0 \tag{9.1}$$

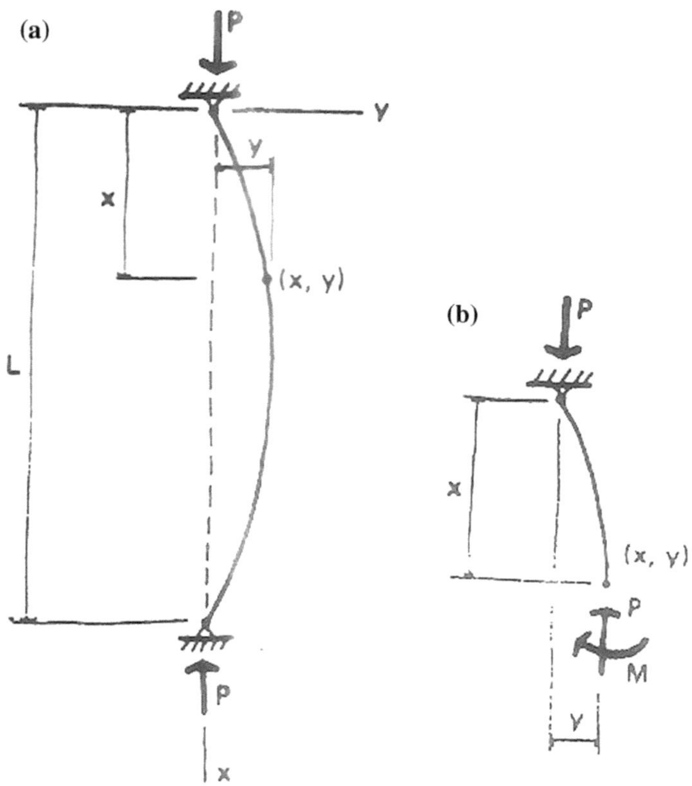

Fig. 9.9 Forces involved in column buckling

If we let $m = \sqrt{P/EI}$ such that:

$$\frac{d^2y}{dx^2} + m^2y = 0 \tag{9.2}$$

The solution of the differential equation is of the form:

$$y = A\cos mx + B\sin mx \tag{9.3}$$

which involves two arbitrary constants, A and B. To evaluate the constants A and B, we can use the boundary condition of $y = 0$ when $x = 0$ from which we get $A = 0$, and therefore, $y =$ Bsin mx. Likewise, $y = 0$ when $x = 1$, or Bsin mL = 0. If B is to have a value, then sin ML must be zero. This is only true if mL = 0, π, 2π, 3π, …, nπ. The coefficient of π represents the buckling mode. For the single-value mode of the column, ML = π.

Replacing m by $\sqrt{P/EI}$ and solving for P, we get:

$$\boxed{\mathbf{P_{cr} = \frac{\pi^2 EI}{L^2}}} \tag{9.4}$$

where P_{cr} is the critical buckling load.

The Euler equation is more useful if we make a minor modification using the radius of gyration concept i.e. $r = \sqrt{I/A}$

r = radius of gyration of the column cross-section (mm) and
A = area of the column cross-section.

If we solve this expression for I and substitute into equation (iv), we have

$$\frac{\mathbf{P}}{\mathbf{A_{cr}}} = \frac{\boldsymbol{\pi^2 E}}{(\mathbf{L/r})^2}$$

where $\left(\frac{P}{A}\right)_{cr}$ = stress caused by the critical loadL/r = slenderness ratioThe critical stress (and load) is inversely proportional to the square of this ratio.

Note the following:

(i) The two equations, i.e. $P_{cr} = \frac{\pi^2 EI}{L^2}$ and $\left(\frac{P}{A}\right)_{cr} = \pi^2 E/(L/r)^2$ can be used interchangeably.
(ii) Neither equation includes any term representing the strength of the material.
(iii) The equations give the failure loads and stresses.
(iv) No factor of safety has been included.

As earlier indicated, the least cross-sectional dimension d_{min} is directly proportional to the radius of gyration for a rectangular cross-sectional wooden column, i.e.

$$\boxed{\mathbf{r_{min} \propto d_{min}}}$$

Hence, the Euler critical buckling stress equation can be rewritten as follows:

$$(P/A)_{cr} = \sigma_{cr} = \frac{\pi^2 E}{(L_e/r_{min})^2} \text{ for square or circular columns}$$

OR

$$\sigma_{cr} = \frac{\pi^2 E}{(L_e/d_{min})^2} \text{ for rectangular columns}$$

where L_e = effective unbraced length of the column (to be discussed later on).

The Euler bucking stress equation was based on the following assumptions:

(i) The column material is homogenous and isotropic.
(ii) The column material is initially straight when load is applied.
(iii) The load passes through the centroid of the material and is truly axial.

The Euler column buckling equation was developed specifically for long slender columns subject to buckling and bending rather than crushing and is suitable for that condition only. For short columns and small values of L/d, the Euler equation cannot be used since σ_{cr} approaches infinity when L/d approaches zero. The underlying assumptions also have other limitations. For example, there can be no perfectly axial loading, and secondly, theoretically, there is no limit to the magnitude of L/d which may be applicable, say one hundred or more, although modern recommended practice limits this ratio, for practical reasons and realistic load values. The equation also has limitations that are specific to wooden column design in the sense that:

- Wood is heterogeneous (not homogenous).
- Wood is autotrophic (anisotropic).
- Wood is a natural material. It is rare to find a piece of wood that is perfectly straight. Besides, it tends to bowl during drying.

In view of the above limitations, the Euler column buckling equation is usually modified in its application to the design of wooden columns. For use in allowable stress design, the modified Euler critical buckling stress equation is expressed as:

$$\boxed{\boldsymbol{\sigma_{cr}} \leq \frac{\mathbf{K_{CE}E}}{(\mathbf{L_e/r_{min}})^2} \quad \text{for square, circular or other shapes}}$$

or

$$\boxed{\sigma_{\mathbf{cr}} \leq \frac{\mathbf{K_{CE}E}}{(\mathbf{L_e/d_{min}})^2} \quad \text{for rectangular columns}}$$

The K_{CE} term incorporates π^2 divided by the factor of safety to convert the Euler equation to:

$$\boxed{\mathbf{P/A} \leq \frac{3.619\mathbf{E}}{(\mathbf{L_e/r_{min}})^2}} \quad \text{For square, circular or other shapes}$$

or

$$\boxed{\mathbf{P/A} \frac{\leq 0.30\mathbf{E}}{(\mathbf{L_e/d_{min}})^2}} \quad \text{For visually graded sawn lumber rectangular columns}$$

For a rectangular lumber product with less variability, such as machine stress-rated lumber (MSRL), a combined reduction factor including a factor of safety is used leading to the following converted Euler equation:

$$\boxed{\mathbf{P/A} \leq \frac{0.418\mathrm{E}}{(\mathbf{l_e}/\mathbf{d_{min}})^2}}$$

Over the years, several formulas and methods have been developed for computing the load-carrying capacity (strength) of a wooden column. Many of them were modifications of the Euler critical bucking stress equation, adapted for design purposes. Five of such methods are presented shortly.

9.4 The General Procedure for Designing a Wooden Column

The general procedure for selecting the size of a wooden column that is expected to carry a given load (P) is to: (1) determine the effective length of the column based on the assumed end conditions; (2) choose a trial section, i.e. the diameter of circular members, or width and thickness of rectangular or square members; (3) compute the stress on the column, i.e. the load divided by the cross-sectional area (P/A); and (4) compare the ***strength*** (actual or modified compression strength parallel to the grain strength) of the column material with the **stress** on the column and conclude as follows:

- If the ***strength*** $\leq$ ***the*** **stress**, the design is UNSAFE. The design should be repeated from step 2, i.e. choose another section.
- If the ***strength*** is slightly > ***the*** **stress**, the design is SAFE.
- If the ***strength*** is far greater than the **stress**, the design is too conservative even though it may be on the side of safety. The design should be repeated from step 2 with the objective of reducing material input and cost.

The various formulas developed over time for computing the load-carrying capacity of solid wooden columns include the following:

9.4.1 The Rankin-Gordon Formula

This empirical formula was developed in an attempt to take care of the inherent deficiencies in the Euler formula, though the formula itself incorporates the Euler equation. It is given as follows:

$$\frac{1}{\sigma_{RG}} = \frac{1}{\sigma_p} + \frac{1}{\sigma_E}$$

$$\boldsymbol{\sigma_{RG}} = \frac{\boldsymbol{\sigma_p} \times \boldsymbol{\sigma_E}}{\boldsymbol{\sigma_p} + \boldsymbol{\sigma_E}}$$

The design equation, therefore, is:

$$\boxed{\mathbf{P/A \leq \sigma_{RG}}}$$

where

σ_{RG} = Rankine-Gordon load-carrying capacity (strength);
σ_P = compression strength parallel to grain of the column material;
σ_E = Euler buckling stress.

Although this formula yields more conservative results in column analysis unlike Euler equation that often results in overdesign, it is an empirical formula which may not be universally applicable.

9.4.2 *Perry-Robertson Formula*

Another complicated formula was developed by Perry-Robertson for wooden column strength analysis and design:

$$\sigma_{PR} = \frac{\sigma_p + (m\ + 1)\sigma_E}{2} - \sqrt{(\sigma_p + (m + 1)\sigma_E)/2)^2 - \sigma_{Gp} \times \sigma_E}$$

The formula can be rewritten thus:

$$\sigma_{PR} = Z\ - \sqrt{Z^2 - \sigma_p\sigma_E}$$
$$\text{where } Z = \sigma_p + (m + 1)\sigma_E$$

The design equation, therefore, is:

$$\boxed{\mathbf{P/A \leq \sigma_{PR}}}$$

where

σ_{RG} = Rankine-Gordon load-carrying capacity (strength);
σ_P = Compression strength parallel to grain of the column material;
σ_E = Euler buckling stress.

It can be seen again that the Euler equation features in this very analytical formula that is based on the assumption that wood assumes a curved shape and that the profile of the curvature is the sine curve. The formula tends to yield a slightly more accurate result that the Rankine-Gordon formula. It however suffers from a number of defects, including the fact that it is very cumbersome; the results the equation yields are still not entirely satisfactory; and lastly, the assumption of initial wood curvature is not entirely justified.

9.4.3 The Madison Approach

This method of wooden column analysis and design was developed in the Forest Products Research Laboratory in Madison, Wisconsin, USA several decades ago. The method is based on the classification of wooden columns into three, i.e. ***short***, ***intermediate*** and ***long*** depending on the slenderness ratio (as previously defined). The permissible/allowable load-carrying capacity of the column is then determined based on the slenderness ratio as follows:

- For a short column with a slenderness ratio of $L_e/d_{min} \leq 11$ or $L_e/r_{min} \leq 38$, the load-carrying capacity is based on the actual compression strength parallel to the grain of the column material ($\boldsymbol{\sigma_P}$), i.e.

$$\boxed{\mathbf{P/A \leq \sigma_P}}$$

where

P = load to be sustained by the column (N);
A = cross-sectional area of the column (mm^2);
σ_P = compression strength parallel to grain of the column material ($N/\ mm^2$).
For intermediate columns, the load-carrying capacity is computed based on the compression strength parallel to the grain of the member multiplied by a modification factor as follows:

$$\boxed{\mathbf{P/A \leq \sigma_p x \left[1 - 1/3\,(L_e/k_r.r_{min})^4\right]} \text{ for square or round columns}}$$

$$\boxed{\mathbf{P/A \leq \sigma_P\, x \left[1 - 1/3(l_e/k.d_{min})^4\right]} \quad \text{for rectangular columns}}$$

where

σ_P = tabulated compression strength parallel to grain.

- For long columns, the load-carrying capacity is typically based on the ratio of the modulus of elasticity to the slenderness ratio of the member and not the compression strength parallel to the grain of the material since the column is expected to fail by buckling. Hence, the load-carrying capacity is determined long columns with hinged ends from the following equations:

$$\boxed{\mathbf{P/A \leq \frac{3.619E}{(L_e/r_{min})^2}} \quad \text{for square and round columns}}$$

$$\boxed{\mathbf{P/A \leq \frac{0.3E}{(L_e/d_{min})^2}} \quad \text{for visually graded rectangular columns}}$$

$$\boxed{\mathbf{P/A} \leq \frac{0.418\mathrm{E}}{(\mathbf{L_e/d_{min}})^2} \quad \text{for machine stress - rated rectangular columns}}$$

where E = modulus of elasticity of the column material.

9.4.4 The FAO Formula

Another procedure was developed by the Food and Agriculture Organisation (FAO) a few decades ago for designing axially loaded wooden columns. This involves computing the slenderness ratio of the trial section; selecting a reduction factor, k, from a table based on the slenderness ratio (Table 9.1); and then computing the strength of the column based on the reduction factor.

The design formula based on the FAO approach, therefore, is:

$$\boxed{\mathbf{P/A} \leq \mathbf{k\sigma_p}}$$

where

P = the load the column is expected to carry.
σ_p = compression strength parallel to the grain for the selected timber species.
A = cross-sectional area of the column.

Table 9.1 Reduction factor for wooden columns

Slenderness ratio		Reduction factor (k)
L_e/d_{min}	L_e/r	
2.9	10	1.0
5.8	20	1.0
8.7	30	0.91
11.5	40	0.81
14.4	50	0.72
17.3	60	0.63
20.2	70	0.53
23.0	80	0.44
26.0	90	0.35
28.8	100	0.28
34.6	120	0.20
40.6	140	0.14
46.2	160	0.11
52.0	180	0.40

Source FAO 1986

9.4.5 The Revised Madison Approach

As earlier mentioned, a procedure for wooden column analysis and design was developed in the Forest Products Research Laboratory in Madison, Wisconsin, USA several decades ago, whereby columns were classified as short, intermediate or long, depending on their slenderness factor, L_e/r_{min} or L_e/d_{min}. However, in 1991, a new equation format for predicting wood column behaviour was published in the National Design Specification for Wood Construction, (NDS). Columns were no longer classified as *short*, *intermediate* or *long*. Instead, the behaviour was defined by a single equation for L_e/d ratios ranging from 0 to 50. Columns with L_e/d ratios greater than 50 are not permitted and are assigned an adjusted compressive stress of zero. One exception is that L_e/d_{min} ratios up to 75 are permitted during construction but not in service.

The NDS (1991) governing condition for the analysis and design of timber column is:

$$\boxed{\mathbf{P/A \leq \sigma_C \cdot C_p}}$$

where

C_P = column stability factor defined as follows:

$$C_{P=}\frac{(1+\sigma_E/\sigma_P)}{2C^2}-\sqrt{\frac{[1+\sigma_E/\sigma_P]^2}{[2C]^2}-\frac{\sigma_{CE}/\sigma_P}{C}}$$

where

σ_P = tabulated compression strength parallel to the grain of the member multiplied by appropriate modifications factors, i.e. $=\sigma_P\ (C_D)\ (C_m)\ (C_T)\ (C_F)$;
σ_E = Euler buckling stress, previously defined as $\frac{k_{CE}E}{(l_e/d)^2}$ where
K_{CE} = 0.3 for VSR; and 0.418 for MSR lumber, glulam, etc.
c = Buckling and crushing interaction factor = **0.8** for sawn lumber; **0.85** for round timber piles; **0.90** for glued-laminated timber;
E = tabulated modulus of elasticity.

9.5 Important Points to Note in the Design of Solid Wooden Columns

- Any of the foregoing formulas could be used for wooden column design, and different design codes have adopted different formulas. However, some of the formulas would yield more conservative results than the others.
- The allowable stress for a square column is about 50% greater than that for a round column of equal area. When designing a round column, it is usually permitted to use the allowable stress for a square column of equal area.
- Long columns with pinned (hinged) ends are not designed for slenderness ratios L_e/r_{min} above 173, or L_e/d_{min} above 50. In cases where $L_e/r_{min} > 173$, or $L_e/d_{min} > 50$, the slenderness ratio can be reduced by installing intermediate bracing or by increasing the minimum thickness. Another alternative is to replace the solid timber member with a spaced column.

9.6 Design of Tapered Wooden Columns

Tapered wooden columns are those which exhibit variations in width and thickness uniformly along their lengths. They may be either rectangular or circular in cross-section and may be tapered at one end or both ends. The use of tapered columns in construction can reduce the material cost and the dead weight of the structure. Occasionally, their use also has appreciable architectural and functional advantages. Their common applications include electric poles, foundation piles and columns in buildings as shown in Fig. 9.10a, b. In these cases, the member is assumed to have uniform taper from end to end as shown in Fig. 9.10c. In less common situations, a turned wood member may also have a uniform taper in both directions, i.e. double taper as shown in Fig. 9.10d. A sawn or glued-laminated member may also have a uniform taper in one direction, i.e. single taper.

In tapered columns, the slenderness ratio is determined as the ratio of the effective length to the effective dimension for the strong and weak planes of the column and the larger of the two values is then selected. For tapered columns of rectangular cross-section, the effective dimension, ‘d’, is computed thus:

$$d = d_{min} + (d_{max} - d_{min})\,[a - 0.15\,(1 - d_{min}/d_{max})]$$

where

d_{min} = the minimum dimension for that plane of the column;
d_{max} = the minimum dimension for that plane of the column;
a = constant whose value depends on the support condition of the column, i.e.
a = 0.7 for large end fixed, small end unsupported or simply supported,
a = 0.3 for small end fixed, large end unsupported or simply supported,
a = 0.5 for both ends simply supported, tapered towards one end and

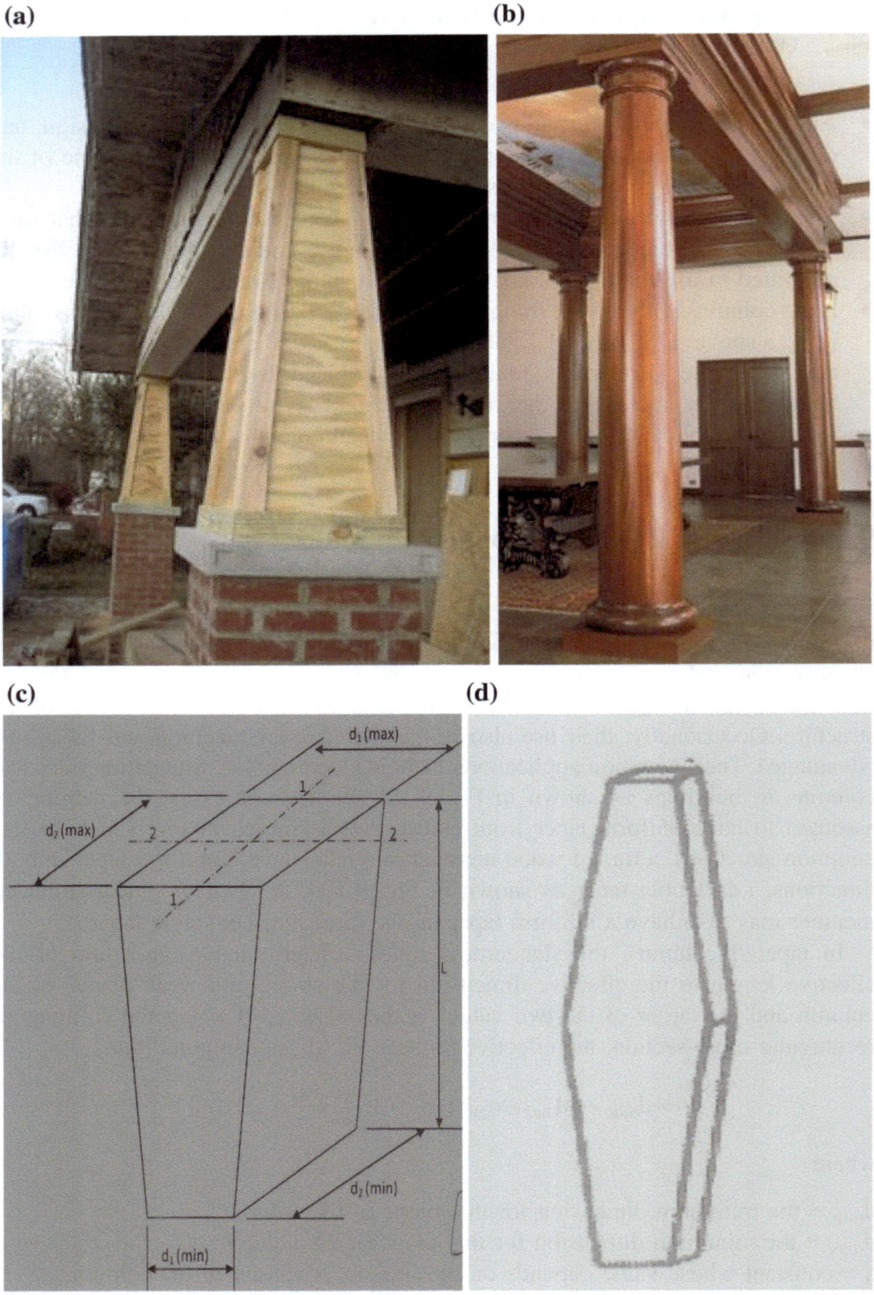

Fig. 9.10 **a** A tapered rectangular column. **b** A tapered circular column. **c** A tapered rectangular column. **d** Double tapered rectangular column

a = 0.7 for both ends simply supported, tapered towards both ends.
For all other support conditions, the effective dimension is taken as

$$d = d_{min} + 0.33(d_{max} - d_{min}).$$

9.7 Pole Buildings

In rural buildings, wood is often used in the form in which it has grown, i.e. round poles. In some areas where enough trees are grown on the farm or in local forests, wooden poles can be obtained at very low cost. These poles have many uses in small building construction such as columns for the load-bearing structure, rafters, trusses and purling. Smaller dimension sticks are often used as wall material or as framework in mud walls. Where straight poles are selected for construction, it will be as easy to work with round timber as with sawn timber. However, somewhat crooked poles can also be used if they are turned and twisted and put into positions in which the effects of the bends are unimportant.

A great number of species can be considered when selecting poles for building construction. Some species are more suitable for silviculture (growing on farms) and silvipasture (growing on pastures) than others, but must always be selected to suit local climatic and soil conditions. Generally, there are several species suitable for each location that are fast and straight growing and produce strong and durable timber. Some species will, in addition to building poles or timber, produce fodder for the animals, fruits, fuelwood. Many species of eucalyptus, from which gum poles are obtained, are very fast and straight-growing hard woods. However, they warp and split easily. Dimensions suitable for building construction are obtained by harvesting the still immature trees. Gum poles provide a strong and durable material if chemically treated.

Tectona grandis, *Casuarina* and several species of *Acacia* are some of the popular tropical hardwoods that produce straight and durable poles. Posts for fencing are obtained by splitting large logs. The posts should be durable, resistant to rot and attack by termites. They are also sometimes used for wall posts in building construction. In coastal areas, mangrove poles are widely used for posts in walls and trusses in roofs.

Pole-type buildings, samples of which are sketched in Fig. 9.11, are common in many parts of Africa where they serve as workshops, farm sheds, etc. Some of the advantages of pole buildings include the following:

- Their construction is generally simpler and less expensive than other types of construction.
- They do not require massive or continuous foundation which can be expensive.
- Site preparation is relatively simple if the construction site is on a level and well-drained terrain.

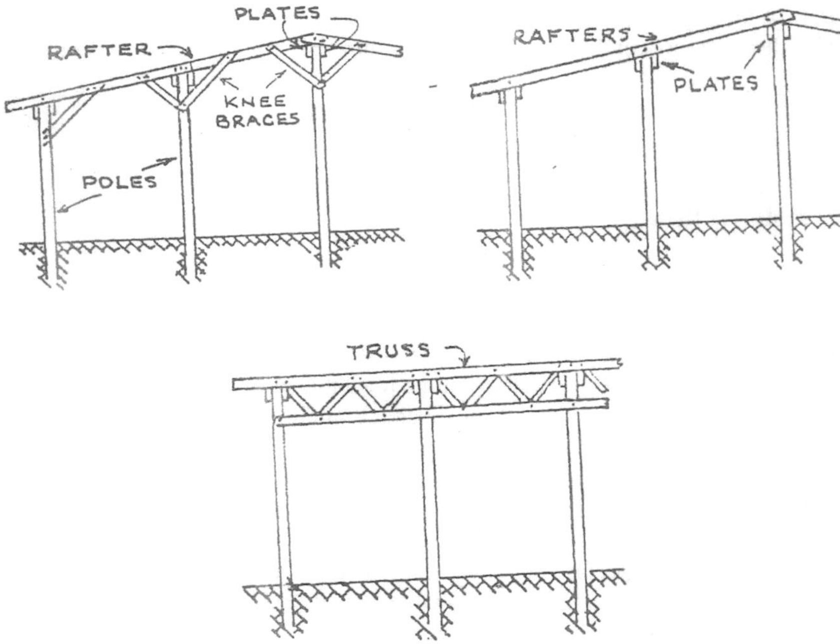

Fig. 9.11 Different types of pole building structure

- A wide variety of building widths and configurations are available.
- Family members of local contractors can often be used for construction.
- Small pole buildings are suitable as do-it-yourself projects.
- Unprocessed roundwood material can be joined by being nailed or tied with string or wire.
- Their construction is in architectural harmony with natural environment.
- They are adaptable to sloping or steep sites.
- There is flexibility for earthquake and high wind design.

Some of the limitations of pole structures include the following:

- Site conditions such as rock outcroppings may limit the depth of poles and may make other foundation types necessary.
- Such buildings are seldom more than one storey high.
- The use of wood posts and trusses often limits their clear spans to 30 m (100 feet) and heights to 6 m (20 feet).

Pole construction has developed into two basic types: pole platform construction in which poles act as a platform for conventional framed construction above floor level; and pole frame construction in which poles act as columns, from the foundations to the roof. The poles in both construction systems can be either embedded into the ground or pinned and supported on concrete footings.

Pole buildings are theoretically assumed to take the form of cantilevers fixed at the lower ends and carrying the wind load perpendicular to the pole axis, while the dead and live loads are carried by the poles acting as columns. In designing the poles, the stress at the point of maximum bending moment is computed using the following governing equation:

$$\sigma_b = \frac{32M}{\pi d^3} \leq \text{Tabulated bending strength/allowable bending stress} \left(N/mm^2\right)$$

where

M = maximum bending moment (N·mm) and
d = diameter at point of maximum bending moment (mm).

The point of maximum bending moment is assumed to occur at 1/4th the depth of embedment below the ground line, unless special restraint at ground level, such as poured concrete slab, is used. For poles with roof bracing, the moment should be checked at the connection to knee braces or the bottom of the roof truss. The maximum compression stress in the pole may be computed using the formula:

$$\text{Compressive strength} = P/A \leq \sigma_E$$

where

P = direct compressive strength (N/mm^2);
A = area of top of pole, or at level where the stress is the greatest (mm^2);
σ_E = allowable compression stress parallel to grain, i.e. Euler buckling stress based on slenderness ratio and elastic modulus of the pole (N/mm^2)

$$\sigma_E = 0.225E/(L/d)^2$$

where

L = unsupported length of pole (mm) and
d = diameter of pole at 1/3 of total length from the top (mm).

The L/d ratio should not exceed 43, and combined axial and bending load based on lateral force and dead load should be checked. The depth of embedment may be calculated using the allowable lateral passive soil pressure determined through soil investigation or other means. A typical value of 11.97 kPa per metre depth is reasonable.

(a)

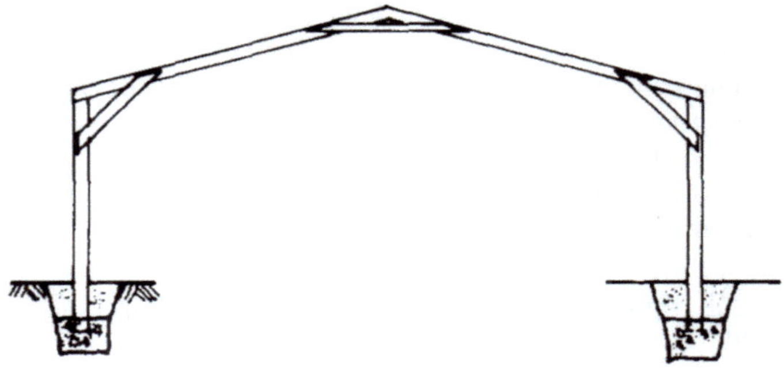

(b)

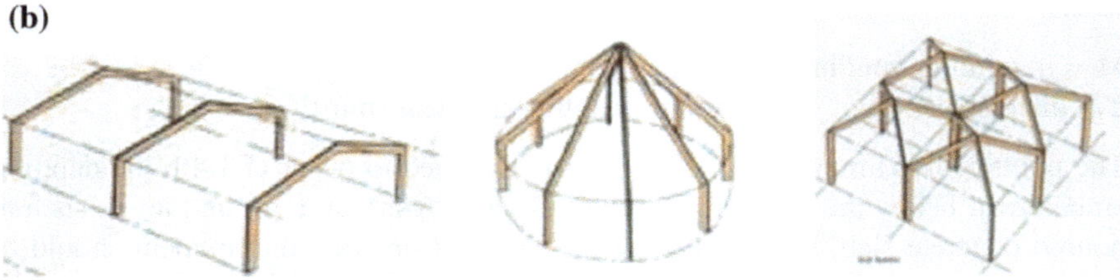

Fig. 9.12 **a** Schematic diagram of a portal frame. **b** Schematic diagram of different forms of timber portal frames

9.8 Timber Portal Frames

Portal frame construction (Fig. 9.12a, b) is a method of designing and building simple structures with timber. The frames may be made from sawn timber, glulam, LVL or plywood webbed beams. The connections between the columns and the rafters are designed to be moment resistant, i.e. they can carry bending forces and must have both strength and stiffness to be effective. Because of these very strong and rigid joints, some of the bending moment in the rafters is transferred to the columns. This means that the size of the rafters can be reduced or the span can be increased for the same size rafters. This makes portal frames a very efficient construction technique to use for wide span buildings such as warehouses and factories, commercial buildings, churches, sporting venues, barns, rural sheds and other places where large, open spaces are required at low cost and a pitched roof is acceptable. They are cost effective, easily accommodate additional lighting and plumbing, and can be erected quickly. The overall appearance is also aesthetically pleasing. They are also particularly suitable for use in potentially corrosive environments such as swimming pool enclosures, buildings housing chemical manufacturer and storage, tanneries. However, portal frames are not wind resistant. They fall over very easily. With spans wider than 20 m, the structural components become massive if timber is used. Hence, steel frames are most common for spans over 20 m.

Generally, portal frames are used for single-storey buildings, but they can be used for low rise buildings with several floors where they can be economic if the floors do not span right across the building. A typical configuration might be where there is office space built against one wall of a warehouse. Timber portal framed buildings may be clad as shown in Fig. 9.13 with conventional metal, or fibre cement cladding, as is common for warehouses and factories, or with timber products such as plywood or weatherboards to provide security and impact resistance. For other commercial applications, cavity brick, brick veneer, concrete tilt-up or concrete block-work may be required for fire isolation or compliance with local building regulations. Where fire ratings are required, the cross-section of glulam portal legs and rafters may be sufficient to provide protection for the members, but extra protection may be required for metal fasteners in the portal knee joint.

For structural design purposes, portal frames are statically indeterminate structures and the complexity of the analysis precludes coverage here. However, the results of such calculations for a number of standard cases of loading are tabulated in handbooks. Using these and the principle of superposition, the designer can determine the structural section required for the frame. Determining the maximum values of the bending moment, shear force and axial force acting anywhere in the frame allow the selection of an adequate section for use throughout the frame. Care must be exercised to ensure that all joints and connections are adequate.

Worked Examples on Solid Wooden Columns

1. Using the six column design formulas, determine the appropriate dimensions for a 1.5 m long N_5, 50% wet rectangular sawn lumber column expected to carry a load of 3 kN if both ends are pinned. Take all modification factors as unity. Assume the following design values:

Fig. 9.13 Cladding of a portal frame

Modulus of elasticity, $E = 6700\ N/mm^2$
Compression parallel to grain (σ_p) = $4.5\ N/mm^2$
Manual Solution
Span (L) = 1500 mm

Expected Load = 3000 N

(i) Compute Effective Length (Le)

$$Ke = 1$$
$$Le = ke \times L = 1500$$

(ii) Select a trial section

$$b = 50\,mm$$
$$d_{min} = 75\,mm$$

(iii) Compute the slenderness ratio

$$L_e/d_{min} = 1500/50 = 30$$
$$(L_e/d_{min})^2 = (30)^2 = 900$$

The stress, P/A = 3000/(50 × 75) = 0.8 N/mm²

(i) **Using the Euler Critical Bucking Stress Equation**

$$\sigma_E \le \frac{0.30E}{(L_e/d_{min})^2}$$

$$\sigma_E = (0.3 \times 6700)/900 = 2.23\,N/mm^2$$

Since $P/A < \sigma_E$, the 50 × 75 mm member is OK and the design is safe.

(ii) **Using the Rankin-Gordon Formula**
The Rankin-Gordon stress, $\sigma_{RG} = \frac{\sigma_G \times \sigma_E}{\sigma_G + \sigma_E}$
The Euler buckling stress σ_E has already been computed above, i.e. $\sigma_E = 2.23\ N/mm^2$

Therefore,
$$\sigma_{RG} = (4.5 \times 2.23)/(4.5 + 2.23) = 2.98\,N/mm^2$$
$$= 1.49\,N/mm^2$$

Since $P/A < \sigma_{RG}$, the 50 × 75 mm member is OK and the design is safe.

(iii) **Using the Perry-Robertson Equation**
Assuming an eccentricity coefficient of 0.85
The Perry-Robertson buckling stress, $\sigma_{PR} = Z - \sqrt{Z^2 - \sigma_p \sigma_E}$

$$Z = \frac{\sigma_p + (m+1)\sigma_E}{2}$$

The Euler buckling stress, $\sigma_E = 2.23\ N/mm^2$
Therefore, $Z = (4.5 + (1.85 \times 2.23))/2 = 4.31\,N/mm^2$

$$\sigma_{PR} = 4.31 - \sqrt{4.31^2 - (4.5 \times 2.23)} = 1.39\,N/mm^2$$

Since $P/A < \sigma_{PR}$, the 50 × 75 mm member is OK and the design is safe.

(iv) **Using the Madison Approach**
The slenderness ratio, $L_e/d_{min} = 1500/50 = 30$
Since $L_e/d_{min} > 11$, the column is either intermediate or long and we have to compute the value of k.

$$K = 0.67\sqrt{E/G} = 0.67\sqrt{6700/4.5} = 25.9$$

Since $L_e/d_{min} > k$, the column is long and has to be designed using the Euler buckling stress equation, i.e.

$$\sigma_E = (0.3 \times 6700)/900 = 2.23\,N/mm^2$$

Since $P/A < \sigma_E$, the 50 × 750 mm member is OK and the design is safe.

(v) **Using the FAO Approach**
The slenderness ratio, $L_e/d_{min} = 1500/50 = 30$
From Table 7.1, reduction factor, k = 0.28 for L_e/d_{min} = 28.8.3, and 0.20 for L_e/d_{min} = 34.6. To obtain the value of k for L_e/d_{min} = 30, we have to perform linear interpolation using the following formula:

$$\frac{y_2 - y}{x_2 - x_1} = \frac{y - y_1}{x - x_1}$$

where

$$y_1 = 0.28,\ y_2 = 0.20,\ x = 30,\ x_1 = 28.3,\ x_2 = 34.6$$

$$\frac{0.20 - 0.28}{34.6 - 28.8} = \frac{y - 0.28}{30 - 28.8}$$

$$y = K_{30} = 0.26$$

The strength of the column = $k\sigma_p = 0.26 \times 4.5 = 1.17\ N/mm^2$
Since $P/A < 1.17\ N/mm^2$, the 50 × 75 mm member is OK and the design is safe.

(a) **Using the Revised Madison Approach**

$$Cp = Z - \sqrt{\frac{Z^2 - \sigma_{CE}/\sigma_P}{C}}$$

where $Z = \frac{(1 + \sigma_E/\sigma_P)}{2C}$

$$C = 0.8$$

$$K_{CE} = 0.3(\text{sawn lumber})$$

Compute C_p

$$l_e/d_{min} = 30 <$$

$$\sigma_E = \frac{K_{CE}E}{(l_e/d)^2}$$

:- $\sigma_E = \frac{0.3 \times 6700}{(30)^2} = \underline{2.23\ \text{N/mm}^2}$

$$\sigma_E/\sigma_C = \frac{2.23}{4.5} = 0.5,\ Z = 1.5/1.6 = 0.94$$

$$\frac{\sigma_E/\sigma_C}{C} = 0.5/0.8 = 0.625$$

$$\begin{aligned} C_p &= 0.94 - \sqrt{(0.94)^2 - 0.625} \\ &= 0.94 - 0.26 = 0.68 \\ \sigma &= 4.5 \times 0.68 = \mathbf{3.06\ N/mm^2} \end{aligned}$$

Since P/A < σ, the 50 × 75 mm member is OK and the design is safe.

2. Using the Rankine-Gordon formula programmed on Excel, determine the appropriate dimensions for a 1.5 m long N_5, 50% wet rectangular sawn lumber column expected to carry a load of 3KN if both ends are pinned. Take all modification factors as unity. Assume the following design values:
   MOE = 6700 N/mm^2
   Compression parallel to grain (σ_p) = 4.5 N/mm^2
   Solution using Excel
   The design computations based on the Rankin-Gordon formula are shown in Fig. 9.14a, while the results are shown in Fig. 9.14b.

Example 3

Write a FORTRAN program based on the modified Madison approach for designing solid wooden columns.

(a)

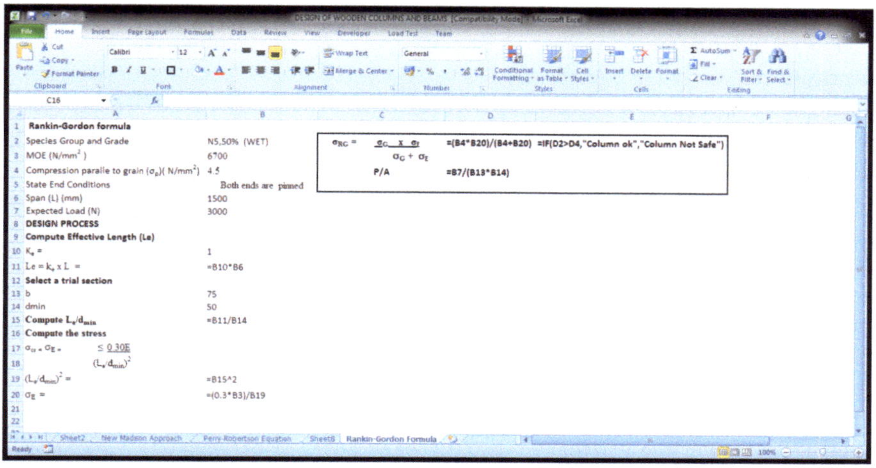

(b)

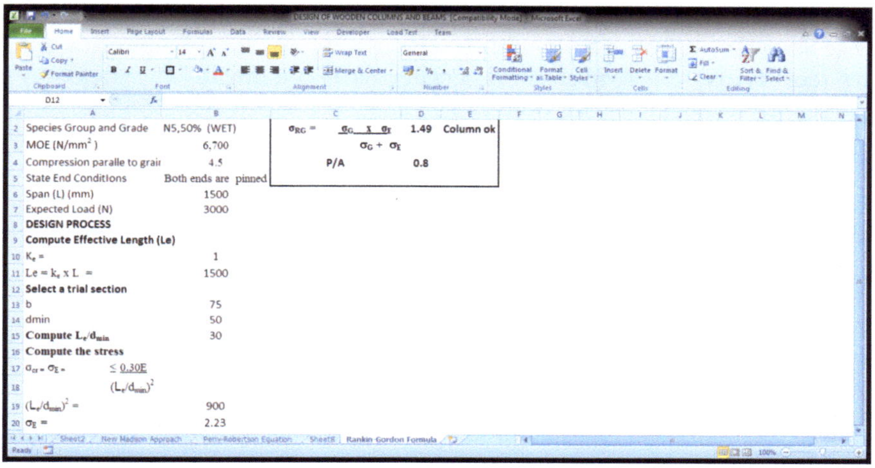

Fig. 9.14 **a** Excel worksheet showing the design computations. **b** Excel worksheet showing the design result

Solution

```
! FORTRAN PROGRAM TO DESIGN WOODEN COLUMNSREAL::CONDITION
WRITE(*,*)'SIMPLE PROGRAM TO DESIGN RECTANGULAR COLUMNS'
WRITE(*,*) "BOTH ENDS FIXED,NO SIDE SWAY, CONDITION = 1"
WRITE(*,*) "ONE END PINNED THE OTHER FIXED, NO SIDE SWAY,CONDITION = 2"
WRITE(*,*) "BOTH ENDS PINNED, NO SIDE SWAY, CONDITION = 3"
WRITE(*,*) "BOTH ENDS FIXED, SIDE SWAY ALLOWED, CONDITION = 4"
WRITE(*,*) "ONE END FIXED, THE OTHER FREE, SIDE SWAY ALLOWED,CONDITION = 5"
```

```
WRITE(*,*) "ONE END FIXED THE OTHER PINNED, SIDE SWAY ALLOWED,CONDITION = 6"
WRITE(*,*) 'CONDITION = '
READ(*,*)CONDITION
WRITE(*,*)
'Input the Compressive Strength parallel to grain of the lumber in Newton
per square mm'
READ(*,*)P
WRITE(*,*)'Input Modulus of Elasticity of the species in Newton per square mm'
READ(*,*)R
WRITE(*,*)'Input the Expected Load on the Column in Newton'
READ(*,*)NL
WRITE(*,*)'Input Length of the Column in mm'
READ(*,*)T
WRITE(*,*)'Input the Preferred thickness on the column in mm'
READ(*,*)V
WRITE(*,*)'Input the Preferred Width (Least Lateral Dimension)
of the column in mm'
READ(*,*)B
IF (CONDITION <=1) GOTO 5
 IF (CONDITION <=2) GOTO 10    IF (CONDITION <=3) GOTO 20
         IF (CONDITION <=4) GOTO 30
           IF (CONDITION <=5) GOTO 40
                        IF (CONDITION <=6) GOTO 50
5 EL = 0.5*T
10 EL = 0.7*T
20 EL = 1.0*T
30 EL = 1.0*T
40 EL = 2.0*T
50 EL = 2.0*T
IF (EL/B > 50) GOTO 180
!C COMPUTATIONS
  CE = 0.3
  NCE = (CE*R)/(EL/V)**2
RAT = NCE/P
C = 0.8
D = 1 + RAT
DEF = D/(2*C)
DF = RAT/C
CP = DEF-SQRT((DEF**2)-DF)
AREA = V*B
STRESS = NL/AREA
STREN = P*CP
   DET = STREN-STRESS
  IF (DET)60,70,80
```

```
   60 GO TO 110
   70 GO TO 150
   80 GO TO 150
      110 WRITE(*,*)'SELECTED  DIMENSONS NOT ACCEPTABLE,
CHOOSE ANOTHER COLUMN CROSS-SECTION'
      180 WRITE(*,*)'SLENDERNESS RATIO OF COLUMN EXCEEDS 50,
CHOOSE ANOTHER COLUMN WIDTH'
             GOTO 200
         150 WRITE(*,*)'SELECTED  DIMENSONS ACCEPTABLE'
                    WRITE(*,*)'STRESS ON THE COLUMN  = ',STRESS
                    WRITE(*,*)'STRENGTH OF THE COLUMN  = ',STREN
         WRITE(*,*)'COLUMN CAN THEREFORE CARRY LOAD  = ',NL
200 END
```

Practice Questions

1. Determine the Euler critical buckling stress and load for a 150 × 150 mm wooden column that is 4.3 m long assume E = 9500 N/mm^2.
2. Rework problem (i) assuming the column has a cross-section of 150 × 175 mm.
3. A platform designed to hold 20 tonnes (20,000 kg) of farm produce is to be supported by ten identical wooden columns manufactured from *milicia excelsa* with a compression parallel to grain strength value of 9.0 N/mm^2 and MOE of 9500 N/mm^2. As a guard against rodent attack on the produce, the platform is to be elevated above the natural ground level to a height of 1.5 m. Design a typical column for the platform using:

   a. The Madison approach,
   b. The FAO approach and
   c. The Euler formula.

4. It is proposed that a load of 300 kN be borne by a set of six wooden columns, each of square cross-section, the columns being equally loaded and acting together (i.e. none can fail independent of the other). Each column is of length 2.3 m, built in at both ends, and restrained form swaying. Using a 50% grade timber with a compression parallel to grain strength value of 11.2 N/mm^2 and MOE of 12,500 N/mm^2, develop an Excel template and use it to determine the smallest dimension permissible for the column, operating as a long column, and the applicable stress.
5. What size of wooden column would be required to support a 6300-kg load when the length is 1.8 m? Assume the column would be fabricated using a 63% grade timber with a compression parallel to grain strength value of 18 N/mm^2 and MOE of 15,000 N/mm^2 and that it would be of rectangular cross-section, pinned at both ends and restrained from swaying.

6. Design the smallest square column to support a concentric compressive load of 72 KN. Assume a 50% lumber with a compression parallel to grain strength value of 9 N/mm^2 and MOE of 10,600 N/mm^2. The unsupported length of the column is 2.4 m.
7. A round column fabricated from timber is 80 mm in diameter and 3.6 m long. What load can it support if the base is fixed against rotation and the top is not restrained from either horizontal movement or rotation? Repeat the design in (b) above assuming that the column is tapered and the diameter at the point one-third of the length of the taper from the small end is 60 mm. Assume a compression parallel to grain strength value of 9 N/mm^2 and MOE of 10,600 N/mm^2.
8. Develop a simple Excel spreadsheet template for designing rectangular wooden columns using the following design approaches:

   a. The Madison approach,
   b. The FAO approach,
   c. The Euler formula and
   d. The modified Madison approach.

10. Develop a simple Excel spreadsheet template for designing circular wooden columns. The using the following design approaches:

   a. The Madison approach,
   b. The FAO approach and
   c. The Euler formula.

11. Develop a Microsoft Excel template for the use in the design of solid wooden columns. The template should have drop-down menu buttons for selecting pre-stored design values.

References

Breyer DE, Fridley KJ, Cobeen KE (1999) Design of wood structures, 4th edn. McGraw-Hill, New York, USA

Faherty KF, Williamson TG (1999) Wood engineering and construction handbook, 3rd edn. McGraw-Hill Handbooks, New York, USA

Goldstein EW (1999) Timber construction for architects and builders. McGraw-Hill Publishers, New York, USA

Gurfinkel G (1973) Wood engineering, published by Southern Forest Products Association, USA

Khurmi RS, Khurmi N (2013) Strength of materials (Mechanics of solids in S.I. units, international edition, S. Chand & Company PVT Ltd., New Delhi, India

Lindley JA, Whitaker JH (1996) Agricultural buildings and structures, revised edition, published by the American Society of Agricultural Engineers, USA

Marshall WT, Nelson HM (1978) Structures, published by Pitman Publishing Ltd., London

Midwest Plan Service (1977) Structures and environment handbook, published b Midwest Plan Service. Ames, Iowa, USA

NCP (2005) Nigerian code of practice for structural timber design, published by the Standards Organisation of Nigeria

Ozelton EC, Baird JA (1976) Timber designers' manual. Granada Publishing Limited, London

Seward D (2003a) Understanding structures, analysis, materials, design, 3rd edn. Palgrave Macmillan, New York, USA

Seward D (2003b) Understanding structures, analysis, materials, design, 3rd edn. Palgrave Macmillan, New York, USA

Shaeffer RE (1999) Elementary structures for architects and builders, 4th edn. Prentice-Hall Publishing Co., New Jersey, USA

Singh S (1982) Strength of materials, 2nd revised edition. Vikas Publishing House PVT Ltd., U.P., India

Willenbrock JH, Manbeck HB, Suchar MG (1998) Residential building design and construction. Prentice Hall, New Jersey, USA

Wolfe R (1998) Structural use of small-diameter round timbers, USDA Forest Service, Forest Products Laboratory (FPL) publication No. 72, Vol 3, Madison, Wisconsin, USA

Chapter 10
Design of Built-Up Wooden Columns

10.1 Spaced Columns

Spaced columns are formed of two or more individual solid members (rectangular columns) with their longitudinal axes parallel. They are typically separated at the ends and at one or more intermediate points of their length by blocking pieces called ***spacers***. Joining at the ends is usually by gluing, nailing, bolting, screwing, or the use of split ring or shear plate connectors to ensure that the individual columns act as a single unit. Spaced columns separated by well-connected spacer blocks are typically used as vertical columns for direct support of vertical loads as shown in Fig. 10.1 but often appear as chords, diagonals and vertical struts or other compression members in wooden trusses. The load is usually applied directly to the ends of the compression member and the end spacer block should normally not extend beyond the end of the column. In trusses, however, the end spacer blocks may extend beyond the ends and the load may be applied through them and their connection to the compressed members.

Spaced columns tend to enhance the performance ability of the column. There is also an improvement in the second moment of area.

10.1.1 Spacer and End Block Provisions

A critical part of spaced column design is the end spacer block connection. The end spacer blocks and their connectors resist shear forces netting through the connectors at the column and block interfaces. These forces can be rationally calculated, but the method is tedious. The forces depend on the elastic modulus of the wood, the slenderness factor and width of the column members. They also depend on the end condition 'a' or 'b' of the design as shown in Fig. 10.2 and which shall be discussed shortly. Increasing values of the elastic modulus, slenderness ratio, member

A.O. Olorunnisola, *Design of Structural Elements with Tropical Hardwoods*,
DOI 10.1007/978-3-319-65343-3_10

Fig. 10.1 A Spaced column

width or area, and fixity factor (to be discussed shortly) , all result in increasing values of the anti-slip forces the connectors must bear.

The connection between the middle spacer block and the column members is non-structural, as this merely serves to keep the two pieces together and does not carry any shear force.

The following provisions apply in the design of spaced columns:

(1) The end blockings must not be less than 6 times the least lateral dimension of the individual unit; that is, $l_d > 6\ d_{min}$ where l_d is the thickness of the blocking and d_{min} is the least lateral dimension of the column material. This is necessary to ensure proper transfer of shear force between the abutting faces of the blocking.

(2) The clearance spacing between shafts (l_s) should not be less than three times the least lateral dimension of the individual column, i.e. $l_s \geq 3\ d_{min}$. These blockings should be so located such that l_s/B is limited to 20 or 0.7 times the greater slenderness ratio for the overall column, whichever is lesser (where l_c = distance between the centroid of one spacer to the other).

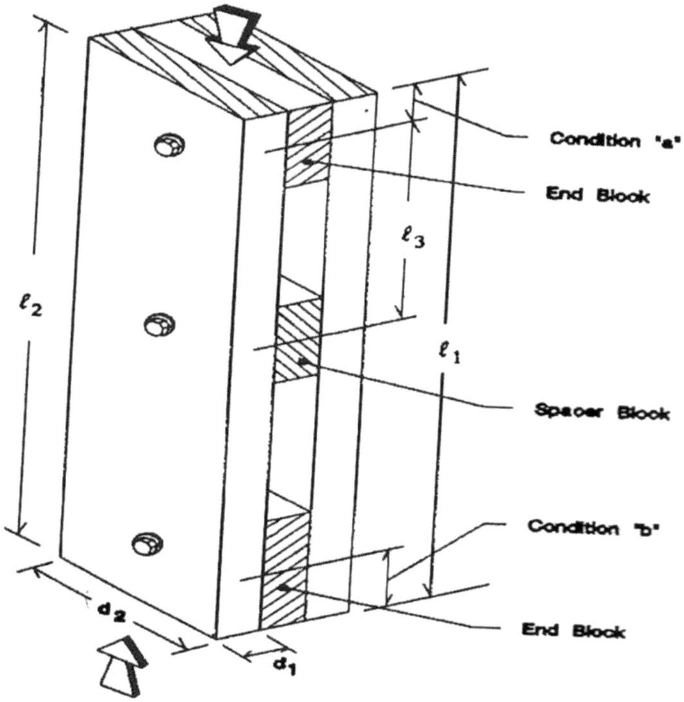

Fig. 10.2 Spaced column Notations

(3) The length of the intermediate blockings (l_d) should be at least 230 mm, i.e. $l_d \geq 230$ mm.
(4) The connectors to be used must be of sufficient strength and rigidity to restrain differential movement between the compression members. The forces causing differential movements are less in short columns than in long spaced columns, and the connector strength requirements are correspondingly less.
(5) In general, the same type of connectors should be used for both end and intermediate blockings. There should be at least 4 screws or 2 bolts per blocking which should be so spaced as to provide even pressure over the area of blocking/packing.
(6) Spacer blocks are required at both ends of the unit. A spacer block is also required at mid-length and should be so fastened as to ensure that the spacer members would maintain their initial spacing. The end and intermediate blockings must be of adequate dimensions to allow the use of nails, screws and adhesives.

(7) There should be at least 2 intermediate blockings/spacers in a spaced column unless the length of the column does not exceed 30 times the least lateral dimension of the unit; that is, L is not greater than 30 B for which condition only a single intermediate pack is required.
(8) The individual members of a spaced column are usually of equal thickness and are designed as single column among which the load is divided proportionately.
(9) Thickness of spacer and end blockings shall not be greater than that of individual members of the spaced column.

10.1.2 Design Criteria for Spaced Columns

The design of spaced columns is similar to that of simple rectangular columns with particular end fixity conditions. However spaced columns are classified as to degree of end fixity, i.e. end condition 'a' or end condition 'b'. The magnitude of the spaced column fixity factor, K_x, is determined by the end condition. If the centroid of the end spacer block connector group is located not more than $L/20$ from the end of the column, a fixity factor, $K_x = 2.5$, is recommended for design. This is called end condition 'a'. If the centroid of the connector group is between $l_1/20$ and $L/10$ from the end of the column, a fixity factor, $K_x = 3.0$, is used. This is called end condition 'b'. In other words,

End condition 'a' applies when the end distance $\leq L/20$; *and* $K_x = 2.5$
End condition 'b' applies when $l_1/20 <$ end distance $\leq L/10$; and $K_x = 3.0$
It is important to note that:

(1) When individual members of a spaced column are of different species grades, or thicknesses, the lesser allowable compression parallel to grain design value for the weaker member shall apply to both members.
(2) The allowable compression parallel to grain design value for a spaced column shall not exceed the allowable compression parallel to grain design value for the individual members evaluated as solid columns without regard to fixity using the columns slenderness ratio.
(3) Spaced column designs should always be checked for strength in their wide face direction as part of a complete design. Usually, they behave as hinged-end columns in that direction, in which use of L_e/d_{min} should not exceed 50.

10.2 Glulam Columns

Glued-laminated timber (glulam) shown in Fig. 10.3 is typically made of materials glued together from smaller pieces of seasoned and accurately surfaced lumber, either in straight or curved form, with the grain of all the laminations essentially parallel to the length of the member. Glue-laminated timber is frequently used in place of sawn timber for truss members, beams, stringers, columns and arches. It

Fig. 10.3 Glulam columns

has the advantage of higher working stresses, being available in larger sizes and longer lengths than sawn timber, and reducing to a minimum of the problems of shrinkage and the resultant secondary stresses in joints.

Glulam members may be either horizontally or vertically laminated. A horizontally laminated member is one comprising four or more laminations and in which the loads on the member act in a plane perpendicular to the plane of the laminations. It is the most commonly used form. A vertically laminated member is one comprising three or more laminations and in which the loads on the member act in a plane parallel to the plane of the laminations. The laminates can be spliced and glued together in such a way as to produce glulam of practically any size and length, the length being limited by factors such as handling systems and length restrictions imposed by highway transport systems, rather than by the size of the tree.

In the fabrication of a glulam member, different wood species may be used. However, it is important that their dimensional changes with changes in moisture content are similar. To avoid as much differential change as possible, the moisture contents at fabrication time of the various laminations should be within a few percent of each other, even when wood of the same species is used. Laminating standards also require that laminations shall not exceed 50 mm (2 inches) in thickness; usually, 16.75 mm (¾ inches) and 37.5 mm (1½ inches) actual thicknesses are employed. The 18.75 mm size is generally used for sharply curved members. The laminations can be any length as required.

Industrial fabrication of glulam consists of various well-defined steps that include selection and preparation of lumber for laminating and layout of laminated assemblies and gluing of the finished member. Glue selection plays an important role in the fabrication of these members. Glues are generally chosen on the basis of their durability under service conditions. Hence, designers must specify to fabricators whether dry or wet use is intended for the given glulam. In the past, only glues of natural origin including animal, vegetable-starch, casein, vegetable-protein and blood-albumin types were used. Casein was the most widely used, and fabricated members were mostly limited to interior used. In modern days, synthetic resin glues including resorcinol resins and phenol–resorcinol combinations are in frequent use. This allows the use of glulam for exterior applications in bridges and utility poles and results in improved performance. Urea adhesives are not permitted for use in structural glulam timber.

10.2.1 Advantages of Glulam Over Solid-Sawn Wood Construction

The following are significant advantages glulam timber offers:

(1) Higher utilization of wood, since lower-grade material can be used for the less highly stressed central laminations in beams without adversely affecting structural integrity. Also, since fabrication of large glulam members is possible from smaller pieces, lumber from smaller trees can be used effectively.
(2) Size and length of the wood members are not limited by length of tree.
(3) Strength-reducing characteristics of wood can be controlled through quality specification of individual laminations. Since the latter are thin enough to be readily seasoned before fabrication, seasoning checks (otherwise associated with large solid-sawn members) are minimized.

(4) For dry conditions of use of the finished member, fully seasoned laminations may be used. Thus, the increased strength of seasoned wood can be realized in design.
(5) Members of variable depth can be designed and fabricated. Economy can be obtained for variations in cross-section that reflect strength requirements.
(6) Architecturally pleasing effects are obtained with glulam design. The laminating process is versatile, enabling the designer to obtain a variety of physical shapes that are not possible with solid-sawn timber.
(7) Laminated timber meeting the size requirements of Heavy Timber Construction receives the fire ratings of that particular category of materials. This is considerably better than light frame wood construction and unprotected steel. Heavy timber structures rarely are consumed by fire. They support their structural loads under extremely intense thermal exposure, permitting safe access to the structure by firemen in establishing control and suppression.
(8) The adhesives used in laminating are not combustible, nor do they melt, soften or lose their strength under the effects of high temperature. The insulating effect of wood further protects the glue lines and the wood on the interior of the members.
(9) Glulam is normally manufactured in the range of 10–16% average moisture content. By specifying moisture content within ±3% of the average service conditions, the designer can avoid dimensional change problems.

10.2.2 Disadvantages of Glulam

(1) Compared to solid-sawn members of equivalent capacity, glulam may be more expensive (about 3–4 times more expensive) because it takes more time to cut and season lumber, and to add the necessary laminating and gluing. Also, the laminating process requires special equipment, plant facilities, adhesives, fabricating skills and quality control that are not needed to produce solid-sawn timbers.
(2) Glulam construction does not compete economically with solid-sawn designs for short spans and small size members.

10.2.3 Design Criteria for Glulam Columns

The design of glulam columns is similar to that of sawn lumber columns. For illustration, the Modified Madison Approach shall be used.

Worked Examples

1. A laminated beam is to be supported by a spaced column at one end where the load on the column will be 27 kN (normal). The length of the column will be 3.6 m. The species is *Milicia excelsa* of 50% grade for the column members which are to be normal 75-mm lumber pieces. Design the spaced column for dry use. To obtain the stiffest column, plan to use end condition 'b'. Assume the following design values: modulus of elasticity, E = 10600 N/mm^2; compression strength parallel to grain, σ = 9 N/mm^2.

Solution

Design Assumptions

(a) pin–pin end condition
(b) 75 × 100 rectangular member

i. **Using the Madison Approach**
Step 1: Determine the effective length of the column based on the assumed end conditions.

(a) $L_e = L$ = 3600 mm for pin–pin end condition

Step 2: Determine the slenderness ratio. Thereafter, check whether the column is intermediate or long, i.e. compare the value of K with the slenderness ratio:

$$L_e/d_{min} = 3600/75 = 48 \text{ which is greater than } 11$$
$$K = 0.671\sqrt{10600/9} = 23.03$$

Since $L_e/d_{min} > K$, it is a long column, therefore it should be designed as a long column, i.e.

$$P/A \leq \sigma_P$$

Step 3: Compute allowable load based on the class of the column

$$\text{Load per column, } P = (27000 \times 0.5) = 13500\,\text{N}$$
$$\text{Stress in each column, } P/A = 13500/7500 = 1.8\ \text{N/mm}^2$$
$$\text{Load carrying capacity of the column, } \sigma_P = \frac{0.3K_xE}{(L_e/d_{min})^2}$$
$$= (0.3 \times 3 \times 10600)/(48)^2 = 4.14\ \text{N/mm}^2$$

Since $P/A < \sigma_P$, 2 No. 75 × 100 mm pieces are ok for fabricating the spaced column

ii. **Using the Revised Madison Approach**
Step 1: State governing equation

$$P/A \leq C_P\sigma_C$$

Recall that P/A in this case is 1.8 N/mm^2
Step 2: **Determine the effective length**
The effective column length for a spaced column shall be determined in accordance with good engineering practice. Actual column length is multiplied by the appropriate buckling length coefficient to determine effective column length, $l_e = (k_e)(L)$, except that the effective column length, l_e should not be less than the actual column length, L.
Step 3: **Evaluate the column stability factor C_p**
The column stability factor is calculated as follows:

$$C_P = \frac{(1+\sigma_E/\sigma_P)}{2C} - \frac{\sqrt{[1+\sigma_E\backslash\sigma_P]^2}}{\sqrt{[2C]^2}} - \frac{\sigma_E/\sigma_P}{C}$$

$$= Z - \sqrt{Z^2} - \frac{\sigma_E/\sigma_P}{C}$$

where $Z = (\frac{1+\sigma_E/\sigma_P}{2C})$
$\sigma_E = \frac{0.3K_xE}{(L_e/d_{min})^2} = 4.14\ \text{N/mm}^2$ (as already calculated using the Madison approach)
All the constants and variables are as previously defined, but

$$K_x; = 2.5 \text{ for fixity condition `a'}$$
$$K_x; = 3.0 \text{ for fixity condition `b'}$$

$$\sigma_E/\sigma_C = 4.14/9 = 0.46$$
$$Z = (1+0.46)/1.6 = 0.91$$
$$\frac{\sigma_E/\sigma_P}{C} = 0.46/0.8 = 0.57$$
$$C_p = 0.91 - \sqrt{(0.91)^2 - 0.57}$$
$$= 0.94 - 0.51 = 0.43$$

Step 4: **Compute Allowable Stress**

$$\sigma = C_P \times \sigma_P = 0.43 \times 9 = 3.87\ \text{N/mm}^2$$

Since P/A < σ, the 75 × 100 rectangular member is OK, and the design is safe

2. The roof of the entrance to a residential building is to be supported at one end with a 2.5-m-long spaced column (Fig. 10.3), fabricated with 50 × 100 mm N_2, 80% timber species in dry service condition. If the total roof

load is approximately 40 kN but only about 25% of the load is borne by the spaced column, check the adequacy of the spaced column using the Madison Approach. Assume the following design values: modulus of elasticity, $E = 12500\ \text{N/mm}^2$,
$\sigma = 18\ \text{N/mm}^2$. State other design assumptions.

Solution

Design Assumptions:
Pin–pin column end conditions, i.e. effective length = actual length = 2500 mm
End condition 'b': $l_1/20 <$ end distance $\leq L/10$; $K_x = 3.0$
Load on spaced column = 0.25 × 40 kN = 10 kN
Load on each of the 2 spaced columns = 10 kN/2 = 5 kN
Column dimensions = 50 × 100 mm
Slenderness ratio = L/d_{min} = 2500/50 = 50−long column
Governing equation for the design: $P/A \leq \boldsymbol{\sigma_P}$

$$\mathbf{P/A}; = 5000\ \text{N}/\left(50 \times 100\ \text{mm}^2\right) = 1\ \text{N/mm}^2$$

$$\sigma_{\mathbf{P}}; = \frac{0.3 K_x E}{\left(L_e/d_{min}\right)^2}$$

$$; = (0.3 \times 3 \times 12500)/(50)^2 = 4.5\ \text{N/mm}^2$$

Since $P/A < \sigma_P$, the 2 No. 50 × 100 mm pieces are OK for the spaced column.

3. Write a FORTRAN program for designing spaced columns

Solution

```
! FORTRAN PROGRAM TO DESIGN SPACED COLUMNS
REAL::CONDITION
WRITE(*,*)'SIMPLE PROGRAM FOR DESIGNING SPACED COLUMNS'
WRITE(*,*) "BOTH ENDS FIXED,NO SIDE SWAY, CONDITION = 1"
WRITE(*,*) "ONE END PINNED THE OTHER FIXED, NO SIDE SWAY,CONDITION = 2"
WRITE(*,*) "BOTH ENDS PINNED, NO SIDE SWAY, CONDITION = 3"
WRITE(*,*) "BOTH ENDS FIXED, SIDE SWAY ALLOWED, CONDITION = 4"
WRITE(*,*) "ONE END FIXED, THE OTHER FREE, SIDE SWAY ALLOWED,CONDITION = 5"
WRITE(*,*) "ONE END FIXED THE OTHER PINNED, SIDE SWAY ALLOWED,CONDITION = 6"
1 WRITE(*,*) 'CONDITION='
READ(*,*)CONDITION
WRITE(*,*) 'INPUT 1 for VISUALLY GRADED OR 2 for MACHINE GRADED Lumber '
READ(*,*)J
WRITE(*,*)'Input the Compressive Strength parallel to grain of the lumber in
Newton/square mm'
READ(*,*)COMPAR
```

```
WRITE(*,*)'Input Modulus of Elasticity of the species in Newton/square mm'
READ(*,*)MOE
WRITE(*,*)'Input the Expected TOTAL LOAD on the spaced Column in Newton'
READ(*,*)TLOAD
WRITE(*,*)'Input Length of the Column in mm'
READ(*,*)LENGTH
WRITE(*,*)'Input the Preferred thickness of each member of the spaced column
in mm'
READ(*,*)THICKNS
WRITE(*,*)'Input the Preferred Width (Least Lateral Dimension) of each
member of the spaced column in mm'
READ(*,*)WIDTH
IF (CONDITION<=1) GOTO 5
IF (CONDITION<=2) GOTO 10
IF (CONDITION<=3) GOTO 20
IF (CONDITION<=4) GOTO 30
IF (CONDITION<=5) GOTO 40
IF (CONDITION<=6) GOTO 50
5 EFLENGTH = 0.5*LENGTH
10 EFLENGTH= 0.7*LENGTH
20 EFLENGTH= 1.0*LENGTH
30 EFLENGTH= 1.0*LENGTH
40 EFLENGTH= 2.0*LENGTH
50 EFLENGTH= 2.0*LENGTH
IF (EFLENGTH/WIDTH>50) GOTO 180
!C COMPUTATIONS
IF (J==1) CEFACTOR=0.3
IF (J==2) CEFACTOR=0.418
NCEFACT= (CEFACTOR*MOE)/(EFLENGTH/THICKNS)**2
RAT=NCEFACT/COMPAR
C = 0.8
D= 1+RAT
DEF= D/(2*C)
DF=RAT/C
CP=DEF-SQRT((DEF**2)-DF)
AREA=THICKNS*WIDTH
STRESS=(TLOAD*0.5)/AREA
STREN=COMPAR*CP
DET=STREN-STRESS
IF (DET)60,70,80
60 GO TO 110
70 GO TO 150
80 GO TO 150
```

```
110 WRITE(*,*)'SELECTED DIMENSONS NOT ACCEPTABLE,CHOOSE ANOTHER COLUMN
CROSS-SECTION'
180 WRITE(*,*)'SLENDERNESS RATIO OF COLUMN EXCEEDS 50,CHOOSE ANOTHER COLUMN
WIDTH'
GOTO 1
150 WRITE(*,*)'SELECTED DIMENSONS ACCEPTABLE'
WRITE(*,*)'STRESS ON EACH COLUMN MEMBER =',STRESS,'N/mm squared'
WRITE(*,*)'STRENGTH OF EACH COLUMN MEMBER=',STREN,'N/mm squared'
WRITE(*,*)'EACH MEMBER OF THE SPACED COLUMN CAN CARRY A LOAD =',
TLOAD*0.5/1000,'KN'
WRITE(*,*)
WRITE(*,*)
WRITE(*,*)'Do you wish to design another spaced column? Please enter 1 for
YES, 2 for NO'
READ(*,*)K
IF(K==1)GOTO 1
IF(K==2)GOTO 200
200 END
```

4. Determine the axial compression load capacity of a 7 m long, 170 × 275 mm glulam column fabricated from 50% grade timber for which the E = 10600 N/mm^2, and σ = 9 N/mm^2. Assume pinned end condition.

Solution

$$
\begin{aligned}
L &= L_e = 7000 \text{ mm} \\
L_e/d_{min} &= 7000/170 = 41.18 \\
\sigma_E &= K_{CE}.E/(L_e/d_{min})^2 \\
K_{CE} &= 0.418 \text{ for glulam} \\
C &= 0.9 \text{ for glulam} \\
\sigma_E &= (0.418 \times 10600)/(41.18)^2 \\
&= 2.6 \text{ N/mm}^2 \\
\sigma_E/\sigma_P &= 2.6/9 = 0.29 \\
C_p &= Z - \sqrt{Z^2} - \frac{\sigma_C/\sigma_P}{C}
\end{aligned}
$$

where

$$Z = (\frac{1 + \sigma_E/\sigma_P}{2C})$$
$$= 1.29/1.8 = 0.72$$
$$\frac{\sigma_C/\sigma_P}{C} = 0.29/0.9 = 0.32$$
$$C_p = 0.72 - \sqrt{(0.72)^2 - 0.32} = 0.28$$

$$\text{The allowable stress}, \sigma = C_P \times \sigma_P$$
$$= 0.28 \times 9 = 2.52\ \text{N/mm}^2$$
$$\text{The allowable load}, P = \sigma \times A$$
$$= 2.52 \times 170 \times 275 = \mathbf{117.8\,kN}$$

5. Assuming that the roof of a residential building is expected to sustain a load is 60 kN and the architect wants use a glulam column fabricated with a timber species whose modulus of elasticity, E = 12500 N/mm^2, and compression strength parallel to grain, σ = 18 N/mm^2. Determine the size of a single glulam that can safely carry 60% of the roof load. Use the Modified Madison Approach.

Solution

Design Assumptions:
Pin–pin column end conditions, i.e. effective length = actual length = 2500 mm
Column dimensions = 75 × 125 mm

Given
E = 12500 N/mm^2, σ = 18 N/mm^2
Load on the glulam column = 0.6 × 60 kN = 36 kN
K_{CE} = 0.418 for glulam
C = 0.9 for glulam

Design Computations

$$\text{Slender ness Ratio} = L/d_{\min} = 2500/75 = 33.3$$
$$\sigma_E = K_{CE}.E\,/(L_e/d_{\min})^2$$
$$\sigma_E = (0.418 \times 12500)/(33.3)^2$$
$$= 4.71\ \text{N/mm}^2$$
$$\sigma_E/\sigma_P = 4.71\,/18\ = 0.26$$
$$C_p = Z - \sqrt{Z^2} - \frac{\sigma_C/\sigma_P}{C}$$

where

$$Z = (\frac{1+\sigma_E/\sigma_\mathbf{P}}{2C})$$
$$= 1.26\,/1.8 = 0.70$$
$$\frac{\sigma_E/\sigma_P}{C} = 0.26/0.9 = 0.29$$
$$C_p = 0.7 - \sqrt{(0.7)^2 - 0.29} = 0.25$$

$$\text{The allowable stress, } \sigma = C_P \times \sigma_P$$
$$= 0.25 \times 18 = 4.5\ \text{N/mm}^2$$
$$\text{The allowable load, } P = \sigma \times A$$
$$= 4.5 \times 75 \times 125 = 4218725\,\text{N}$$

Since the load on the glulam column, 36 kN < 42.2 kN, the 75 × 150 dimension selected is OK

6. Develop a simple FORTRAN programming code for designing GLULAM columns.

Solution

```
REAL::CONDITION
WRITE(*,*) "BOTH ENDS FIXED,NO SIDE SWAY, CONDITION = 1"
WRITE(*,*) "ONE END PINNED THE OTHER FIXED, NO SIDE SWAY,CONDITION = 2"
WRITE(*,*) "BOTH ENDS PINNED, NO SIDE SWAY, CONDITION = 3"
WRITE(*,*) "BOTH ENDS FIXED, SIDE SWAY ALLOWED, CONDITION = 4"
WRITE(*,*) "ONE END FIXED, THE OTHER FREE, SIDE SWAY ALLOWED,CONDITION = 5"
WRITE(*,*) "ONE END FIXED THE OTHER PINNED, SIDE SWAY ALLOWED,CONDITION = 6"
WRITE(*,*) 'CONDITION='
READ(*,*)CONDITION
WRITE(*,*)'Input the Compressive Strength parallel to grain of the lumber in
Newtons per square mm'
READ(*,*)P
WRITE(*,*)'Input Modulus of Elasticity of the species in Newtons per square
mm'
READ(*,*)R
WRITE(*,*)'Input the Expected Load on the Column in Newton'
READ(*,*)NL
WRITE(*,*)'Input Length of the Column in mm'
READ(*,*)T
WRITE(*,*)'Input the Preferred thickness on the column in mm'
READ(*,*)V
```

```
WRITE(*,*)'Input the Preferred Width (Least Lateral Dimension) of the column
in mm'
READ(*,*)B
IF (CONDITION<=1) GOTO 5
IF (CONDITION<=2) GOTO 10
IF (CONDITION<=3) GOTO 20
IF (CONDITION<=4) GOTO 30
IF (CONDITION<=5) GOTO 40
IF (CONDITION<=6) GOTO 50
5 EL = 0.5*T
10 EL= 0.7*T
20 EL= 1.0*T
30 EL= 1.0*T
40 EL= 2.0*T
50 EL= 2.0*T
IF (EL/B>50) GOTO 180
!C COMPUTATIONS
CE=0.418
NCE= (CE*R)/(EL/V)**2
RAT=NCE/P
C = 0.9
D= 1+RAT
DEF= D/(2*C)
DF=RAT/C
CP=DEF-SQRT((DEF**2)-DF)
AREA=V*B
STRESS=NL/AREA
STREN=P*CP
DET=STREN-STRESS
IF (DET)60,70,80
60 GO TO 110
70 GO TO 150
80 GO TO 150
110 WRITE(*,*)'SELECTED DIMENSONS NOT ACCEPTABLE,CHOOSE ANOTHER COLUMN
CROSS-SECTION'
180 WRITE(*,*)'SLENDERNESS RATIO OF COLUMN EXCEEDS 50,CHOOSE ANOTHER COLUMN
WIDTH'
GOTO 200
150 WRITE(*,*)'SELECTED DIMENSONS ACCEPTABLE'
WRITE(*,*)'STRESS ON THE COLUMN =',STRESS
WRITE(*,*)'STRENGTH OF THE COLUMN =',STREN
WRITE(*,*)'COLUMN CAN THEREFORE CARRY LOAD =',NL
200 END
```

Practice Questions

1. Develop a simple Excel template for designing spaced columns.
2. A beam is to be supported by a spaced column at one end where the load on the column will be 3 kN (normal load). The length of the column will be 3.7 m. The species is *Lophira alata* of 63% grade for which the modulus of elasticity, $E = 15{,}000\ \text{N/mm}^2$, and compression parallel to grain, $\sigma = 18\ \text{N/mm}^2$, for beam and column members. Design the spaced column using:

   a. The Madison Approach
   b. The Revised Madison Approach

3. Determine the dimensions of a 10-m-long glulam column. The column is produced from a timber whose modulus of elasticity, $E = 12{,}500\ \text{N/mm}^2$, and compression parallel to grain, $\sigma = 14\ \text{N/mm}^2$. Assume that the axial load to be carried by the column is 500 kN. Assume pinned end condition. Use:

   a. The Madison Approach
   b. The Revised Madison Approach

4. An 80% grade timber ($E = 9{,}000\ \text{N/mm}^2$, and $\sigma = 9\ \text{N/mm}^2$) of nominal size 75×150 mm or wider is to be used for a 42-m-long spaced column, end condition 'a'. Using the Revised Madison Approach, determine the size required to support a load of 36 kN for two-month duration under dry condition.

References

Breyer DE, Fridley KJ, Cobeen KE (1999) Design of wood structures, 4th edn. McGraw-Hill, New York, USA

Faherty KF, Williamson TG (1999) Wood engineering and construction handbook, 3rd edn. McGraw-Hill Handbooks, New York, USA

Goldstein EW (1999) Timber construction for architects and builders. McGraw-Hill Publishers, New York, USA

Gurfinkel G (1973) Wood engineering. Published by Southern Forest Products Association, USA

NCP (2005) Nigerian code of practice for structural timber design. Published by the Standards Organisation of Nigeria

Ozelton EC, Baird JA (1976) Timber designers' manual. Granada Publishing Limited, London

Seward D (2003) Understanding structures, analysis, materials, design, 3rd edn. Palgrave Macmillan, New York, USA

Shaeffer RE (1999) Elementary structures for architects and builders, 4th edn. Prentice-Hall Publishing Co., New Jersey, USA

Chapter 11
Design of Wood Connections

A critical aspect of wood structures is the manner by which members are connected. Wood products are easily drilled, chiselled or otherwise shaped to facilitate the connection of members, and a number of methods as well as a wide range of products are available for connecting wood. The importance of fasteners cannot be overemphasized since wood structures are made up of elements that must be connected together, i.e. every wood structure is a framework composed of parts fitted and joined together.

There are many different types of joint details that can be used in wood connections. When structural members are attached with fasteners or some other type of hardware, the joint is said to be a mechanical connection. A wide array of metal fasteners is available, ranging from the nails and the light connectors used for light framing construction to the bolts, side plates and other hardware used for heavy member connections. When used appropriately, metal fasteners provide means of connection which are easy to install and which offer trouble-free performance. Nailing, for example, is an effective means of connection which, when applied according to specified layouts, results in strong structural systems which perform well even under the most adverse loading conditions such as the effects of earthquake.

Mechanical fasteners are generally designed to transmit axial and shear forces between connected members. However, in some cases, especially when the connection is rigid, they may also be designed to transmit bending loads. Each of the different types of wood fasteners falls into one of two broad categories: those for which holes are not required or are only provided to ease driving and minimize splitting, and those for which holes are required. For example, nails and spikes are do not require holes, while bolts, lag screws, split rings and shear plates cannot be installed without holes. In general, a given fastener's category is a function of its diameter; the smaller a fastener's diameter, the less likely a hole will be needed to

A.O. Olorunnisola, *Design of Structural Elements with Tropical Hardwoods*,
DOI 10.1007/978-3-319-65343-3_11

install it. Regardless of the type of fastener, it must be able to sustain and transfer loads over a large enough area of the wood so that the wood fibre in contact with the fastener is not deformed.

The methods used to join members include lapped and butt connectors. Bolt and connector joints, nailed joints and glued joints and sometimes a combination of two are examples of lapped connections. Butt connections are those that require the use of plates or gussets. In all cases, the joints should be designed by calculating the shear forces that will occur in the members. If two members overlap, the joint is called a single-lap joint. If one is lapped by two other members, i.e. sandwiched between them, it is called a double-lap joint. With a single lap, the joint is under eccentric loading. Sandwich construction enables the necessary sectional area of a member to be obtained by the use of relatively thin timbers. The use of gussets permits members to butt against each other in the same plane, avoids eccentric loading on the joints and provides, where necessary, greater joining area than is possible with lapped members. When full-length timber is not available for a member, a butt joint with cover plates can be used to join two pieces together.

Mechanical connections are different from connections made with adhesives. In general, adhesives are normally used in a controlled environment, such as a glulam, plywood or wood I-joist manufacturing plants under the control of a formal quality assurance programme. The focus of this chapter is largely on mechanical connections and nailed connections in particular.

11.1 Types of Joint Loads

Mechanical connections are generally classified according to the direction of loading. Shear connections, also known as *laterally loaded* connections, are those in which the load is applied perpendicular to the length of the fastener. These connections are further classified as to the number of shear planes. The most common applications are *single shear* and *double shear* (as shown in Figs. 11.1 and 11.2), but additional shear planes are possible. The second major type of loading in a wood connection is one in which the load is applied parallel to the length of the fastener, and the fastener is loaded in tension. This is known as *withdrawal loading* (Fig. 11.3). There are typically two sides to any member (wood, metal or concrete) involved in a mechanical connection. The *point side* or *main member* is the member that receives the pointed end of the fastener, while the *head side* or *side member* is the member that contains the bigger end of the fastener.

Fig. 11.1 Fastener laterally loaded in single shear

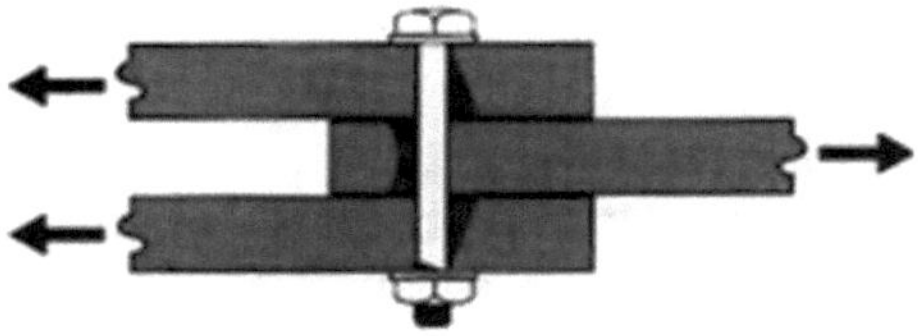

Fig. 11.2 Fastener laterally loaded in double shear

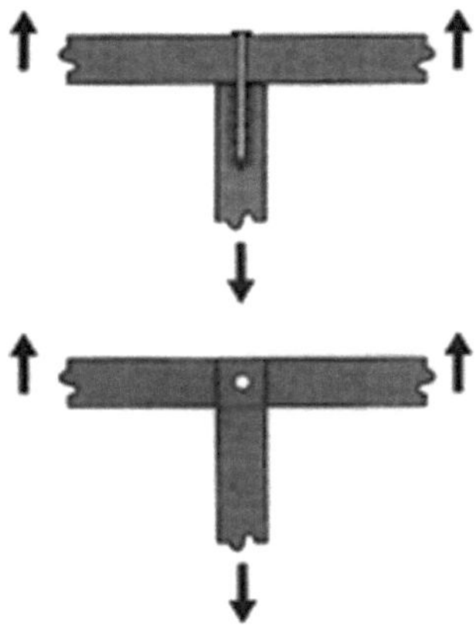

Fig. 11.3 Front and side views of a fastener loaded in withdrawal

11.2 Nailing

Of the various mechanical fasteners used in wood connections, nails, bolts and lag bolts (lag screws) are the most common. Other types of the fasteners include connector plates, sheet-metal anchors among others. Nails, wood screws, sheet-metal anchors and connector plates are commonly used for light-frame construction, while bolts, lag screws and timber connectors are commonly used in heavy timber construction. Nailing is the most basic and most commonly used means of attaching members in wood frame construction in tropical Africa given its relative cheapness and wide availability. A nail is made up of three parts: the head, the shank and the nail point. The head is the part that receives the hammer while nailing, the shank is the 'stem' of the nail, while the nail point is the terminal part of the nail that initiates penetration into the wood member. Also, nails are made from different materials and are produced with different types of surface conditions. The different types of nails are, therefore, distinguished by the following characteristics: nail head, shank, nail point, material type and surface condition.

11.2.1 Heads

There are different types of nail heads. Examples include the following:

- **Flat Counter-Sink Head**: For nail concealment; light construction, flooring and interior trim.

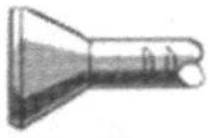

- **Gypsum Wallboard Head**: For gypsum wallboard.

- **Finishing Head**: For nail concealment; cabinetwork, furniture.

- **Flat Head**: For general construction.

- **Large Flat Head**: For tear resistance; roofing paper.

- **Oval Head**: For special effects; cladding and decking.

11.2.2 Shanks

There are also different types of nail shanks. Typical examples include the following:

- **Smooth shank**: For normal holding power; temporary fastening.

- **Spiral or Helical shank**: For greater holding power; permanent fastening.

- **Ringed shank**: For the highest holding power; permanent fastening.

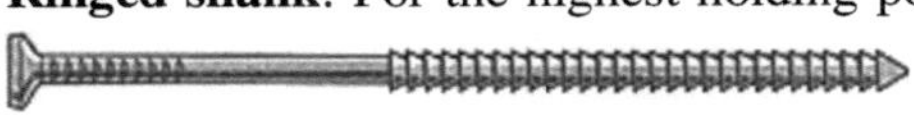

11.2.3 Nail Points

The different types of nail points include the following:

Diamond: For general use, 35° angle; length about 1.5 × diameter.

Blunt Diamond: For harder wood species to reduce splitting, 45° angle.

Long Diamond: For fast driving, 25° angle; may tend to split harder species.

Duckbill: For ease of clinching.

Conical: For use in masonry; penetrates better than diamond.

11.2.4 Materials

The types of materials used for nails are shown in Table 11.1. The most common nails are made of low- or medium-carbon steels or aluminium. Medium-carbon steels are sometimes hardened by heat treating and quenching to increase toughness. Nails of copper, brass, bronze, stainless steel and other special metals are

Table 11.1 Nail materials, finishes and coatings

Material	Application
Aluminium	For improved appearance and long life: increased strain and corrosion resistance
Steel—mild	For general construction
Steel—high carbon	For special driving conditions: improved impact resistance
Stainless steel, copper and silicon bronze	For superior corrosion resistance: more expensive than hot-dip galvanizing
Finishes and Coatings	
Bright	For general construction, normal finish, not recommended for exposure to weather
Blued	For increased holding power in hardwood, thin oxide finish produced by heat treatment
Heat-treated	For increased stiffness and holding power: black oxide finish
Phosphate-coated	For increased holding power; not corrosion resistant
Electro-galvanized	For limited corrosion resistance; thin zinc plating; smooth surface; for interior use
Hot-dip galvanized	For improved corrosion resistance; thick zinc coating; rough surface; for exterior use

available if specially ordered. Uncoated steel nails used in areas subject to wetting will corrode and result in staining of the wood surface. In addition, naturally occurring extractives in some timber species react with unprotected steel and with copper and electro-galvanized fasteners. In such cases, hot-dip galvanized nails or stainless steel or copper nails should be used.

11.3 Types of Nails

The most commonly used types of nails in wood construction are:

(1) **Common wire nails** are used for framing where there will be considerable lateral load. They represent the base structural nail. The diameter is larger than that of box nails, hence greater design value allows load.

(2) **Box Nails**: These are smaller in diameter than common (wire) nails and are used for installing sheathing or roof decks. Although they have lesser holding power, they also reduce the danger of splitting and have smaller diameter than common nails and hence lower allow load. Have greater tendency to bend during driving.

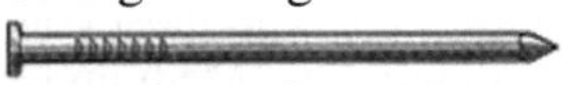

(3) **Common spikes**: These are larger in diameter than common nails, though they have similar lengths. Their larger diameter gives them greater holding power hence they can be used without splitting wood.

(4) **Threaded Hardened-Steel Nails**: These are made from high-carbon steel and are heat-treated and tempered to provide greater strength. They are especially useful when the lumber moisture content will vary during the service life of the structure. The threaded shank provides greater withdrawal capacity in these circumstances. Also, the higher yield strength values associated with high-carbon steel provide improved lateral (shear) resistance. Resistance to withdrawal is increased because of the larger friction between the nail and the wood. However, the withdrawal resistance is in most cases only temporary. Much of the coating may be removed in the driving of the nail.

(5) **Casing Nails**: These are slender low-carbon nails with a casing head and a diamond point. They may be zinc-coated (galvanized to reduce staining of wood exposed to the weather).

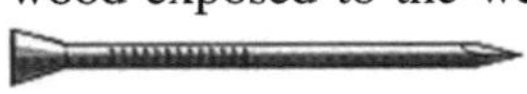

(6) **Cooler Nails**: These are slender, low-carbon steel nails with a flat head and diamond point, and usually cement-coated. The head diameter is the same as or smaller than that of a common wire nail of the same length.
(7) **Zinc-coated (Galvanized) Nails**: These are intended primarily for use where corrosion and staining resistance are important factors in performance and appearance. Also, resistance to withdrawal may be increased if the zinc coating is evenly distributed. However, extreme irregularities of the coating may actually reduce withdrawal resistance.
(8) **Cement-Coated Nails**: These are coated by tumbling or submerging in a resin or shellac (cement is not used).

11.4 Nail Size and Lateral Loading

Nail diameter is specified by gauge number (British Imperial Standard) and is the same as the wire diameter used in manufacture. In the USA, the length of nails is designated by 'penny' or 'pennyweight' abbreviated 'd'. For example, a twenty-penny nail (20d) has typically a length of 100 mm. For a given pennyweight, all the four basic types of nails—common nails, spikes, box nails and threaded hardened-steel nails—have the same length. The first three types have the same basic form, but the diameters are different. Basic nail and spike sizes are presented in Table 11.2. The lateral design values for various sizes of round wire nails driven into the side grain of seasoned wood as shown in Fig. 11.4 are usually tabulated in the codes of practice for timber design.

The allowable load is often based on the assumption that nails are driven at approximately right angles into the side grain of seasoned wood with 19% or less moisture. Also, when more than one nail is used, the sum of the value for single nails may be used. Design values for nails used with kiln-dried, fire-reluctant treated wood may be reduced by 10%. Clinching nails that protrude through joint members may also increase the lateral resistance by as much as 50%. In many cases, it is not possible to nail directly through the side member into the main (holding) member. In these circumstances, toenails may be used. Toenails are nails that are driven at an angle of 30 to the side member and are started approximately one-third of the nail length from the intersection of the two members as shown in Fig. 11.5a. The longitudinal axes of the two members of a toe-nailed joint intersect in the plane of the joint. Also, toenailing can provide joints of greater strength than perpendicular nailing through the side member into the end grain of the main member.

Table 11.2 Nail and spike sizes

Pennyweight	Size designation (Length, mm)	Diameter (mm)			
		Common nail	Box nail	Hardened nail	Spike
2d	25	1.83			
3d	31.25	2.11	1.93		
4d	37.5	2.67	2.03		
5d	43.75	2.67	2.03		
6d	50	2.87	2.51	3.05	–
7d	56.25	2.87	2.51	3.05	–
8d	62.5	3.33	2.87	3.05	–
10d	75	3.76	3.33	3.43	4.88
12d	81.3	3.76	3.76	3.43	5.26
16d	87.5	4.11	3.76	3.76	5.72
20d	100	4.88	4.11	4.50	6.20
30d	112.5	5.26	4.88	4.50	6.68
40d	125	5.72	5.26	4.50	7.19
50d	137.5	6.20	5.72	4.50	7.19
60d	150	6.68	6.20	4.50	–
70d	175	–	–	5.26	–
80d	200	–	–	5.26	–
90d	225	–	–	5.26	–
	175	–	–	5.26	–
	212.5	–	–	9.53	–

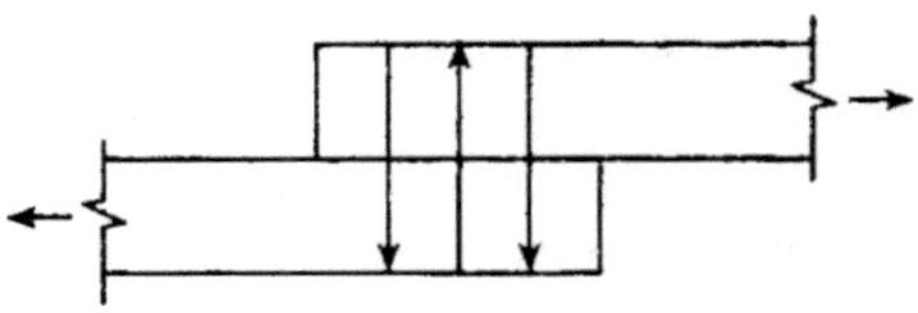

Fig. 11.4 Lateral loading

11.4.1 Penetration

The penetration (p) of a nail is defined as the distance that the nail extends into the main member. It is computed as the length of the nail (L) minus the thickness of the side member (t_s), i.e.

$$p = L - t_s \leq t_m$$

where t_m is the thickness of the main member. If the nail extends beyond the main member, the penetration is taken as the thickness of the main member. Minimum

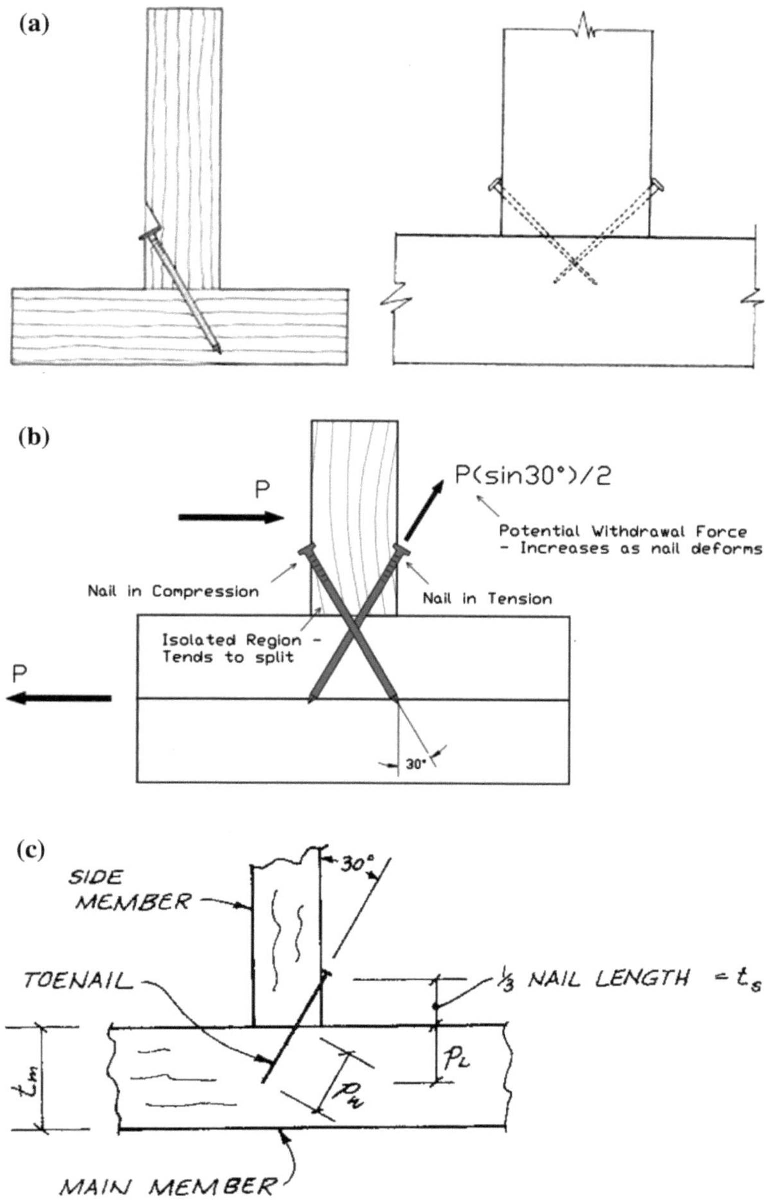

Fig. 11.5 **a** Examples of toe-nailed wood members. **b** Laterally loaded toenails. **c** Toenail penetration in wooden main member

Table 11.3 Nail penetration requirements based on specific gravity of the wood species

Specific gravity	Minimum penetration requirement
0.62 and above	10 × Nail diameter (10D)
0.51–0.55	11 × Nail diameter (11D)
0.42–0.46	13 × Nail diameter (13D)
0.31–0.40	14 × Nail diameter (14D)

penetration requirements for nails are usually specified in the Code of Practice for timber design. These values tend to depend on the specific gravity of the timber materials to be joined. A sample is shown in Table 11.3. The penetration of a toenail for laterally loaded connection (P_L) as shown in Fig. 11.5b, c, which is the projected length of toenail in main member, is given as:

$$P_L = L.\text{Cos}\,30 - t_s \leq tm$$

11.4.2 Spacing

In order to permit the development of the full load at each nail, and to avoid splitting of the wood, minimum spacing between nails and distances from the edges and ends of the member are necessary. Nailing pattern is dependent on the quality and type of nails and timber used, and is based on the safe lateral nail load. Spacing and load arrangement considerations for nails and other metal fasteners (Fig. 11.6) are also usually specified in the Code of Practice for timber design. The ***end distance*** is the distance measured parallel to grain from the square-cut end of a member to the centre of the nearest nail. The ***edge distance*** is the distance measured perpendicular to grain from the edge of a member to the centre of the nearest nail. Other spacing requirements for nailing include ***inter-row spacing*** which is the side spacing between lines of nails; and ***row spacing*** which is the spacing along the grain between adjacent nails.

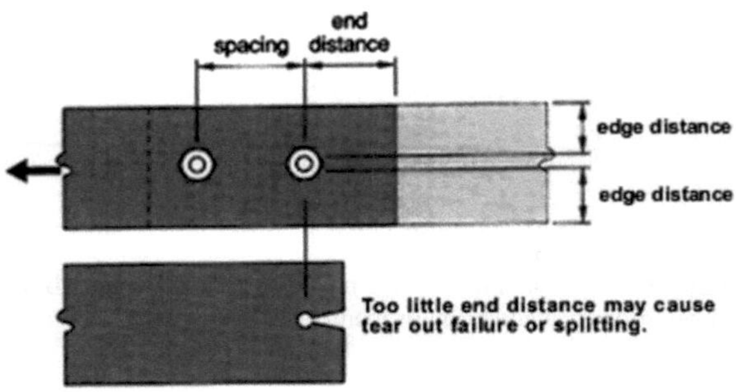

Fig. 11.6 Terminology for metal fasteners spacing

Rules for minimum nail spacing are usually based on nail diameters as shown in Fig. 11.7 and Table 11.4. Nail spacing (Fig. 11.8) should be sufficient to prevent unusual splitting. Nail spacing is also a function of load resistance requirements. Plywood, box beams and other lumber structural elements require specific nailing patterns that can be calculated.

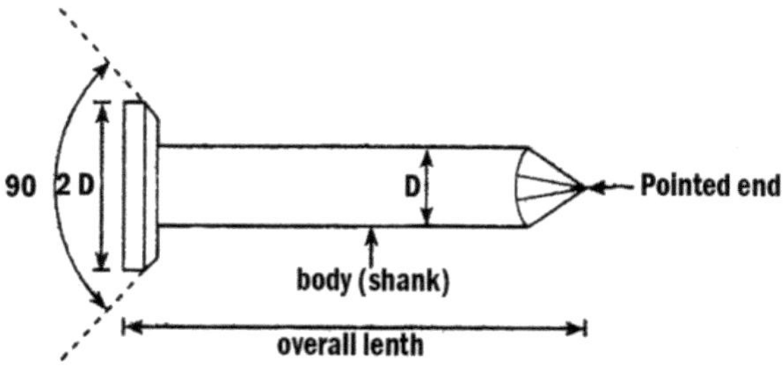

Fig. 11.7 Nail dimensions

Table 11.4 End distance and spacing of nails

Spacing	Driven without pre-boring	Driven into hole pre-bored to 80% of nail diameter
End distance	20D	10D
Edge distance	5D	5D
Inter-row spacing	10D	3D
Row spacing	20D	10D

D = nail diameter

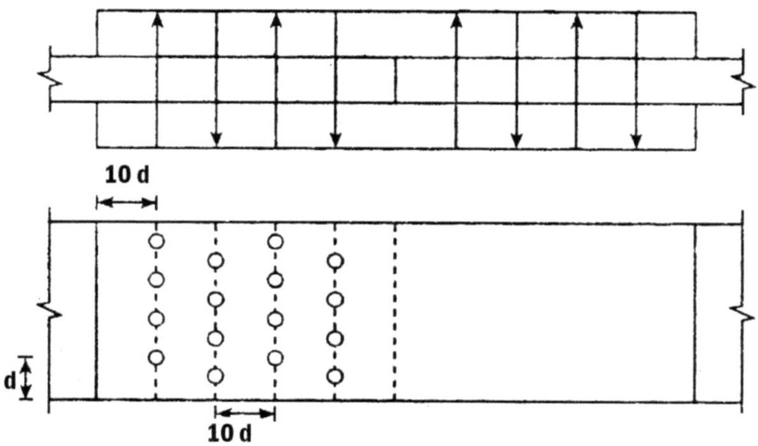

Fig. 11.8 Nail spacing

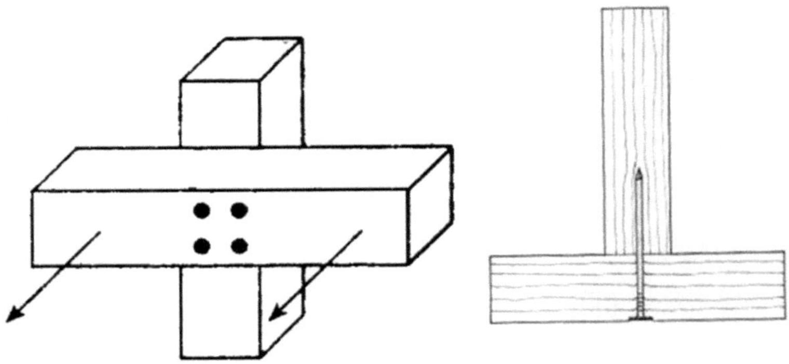

Fig. 11.9 Nails loaded in withdrawal

11.5 Safe Withdrawal Loads for Nails

Examples of nails loaded in withdrawal are shown in Fig. 11.9. Different codes of practice give safe withdrawal design values for various sizes common and threaded steel nails.

The design values usually apply to nails driven into side grain for either seasoned or unseasoned wood. When a nail is driven into wood parallel to the grain, the holding power may be reduced as much as 50%. Hence, nails driven into end grain should not be loaded in withdrawal. Design values for toe-nailed joints may be taken as 67%. Also, values for nails used with kiln-dried, fire-reluctant treated wood should be reduced by 10%. The penetration of toenail for a connection loaded in withdrawal (Pw as previously shown in Fig. 10.13b) which is equal to the total length of nail in main member is given as:

$$\mathrm{Pw} = 1 - \frac{0.33}{\mathrm{Cos}\,30} \leq \frac{\mathrm{tm}}{\mathrm{Cos}\,30}$$

11.6 Guidelines for Creating Nailed Joints

The following are some of the rules of thumb that may be followed in creating effective and efficient nail joints:

- Nail diameter should be between 1/4 and 1/6 times the thickness of the side (thinner) member.
- The length of the nail should be at least 2.5 times the thickness of the side member.

- In lengthening joint, a minimum 8 nails (4 nails on either side) should be used. In node joints, a minimum 2 nails should be used.
- The maximum numbers of nails in one row, in the direction of force, should be 10.
- The minimum thickness of any single member should be 15 mm.
- The maximum thickness of any single member should be 100 mm.
- Width of a member should not exceed eight times its thickness.
- Where a number of nails are used in the joint, they should be so arranged that the centroid of the group lies on the axis of the member.
- The force should pass through a centroid of the nail group.

11.7 Bolted Connections

Most bolts and nuts (Fig. 11.10) are used in laterally loaded connections. When loads on wood connections are particularly heavy, bolted joins are advisable. Most bolted connections are either single-shear (two-member) or double-shear (three-member) joints. Connections can be wood-to-wood or wood-to-metal. A single-shear connection also occurs when a wood member is attached by anchor bolts to a concrete foundation or a concrete or masonry wall. These are called 'wood-to-concrete' or 'wood-to-masonry connections', or simply 'anchor bolt connections'. These connections are shown in Fig. 11.11a–e.

Fig. 11.10 Bolts and nuts

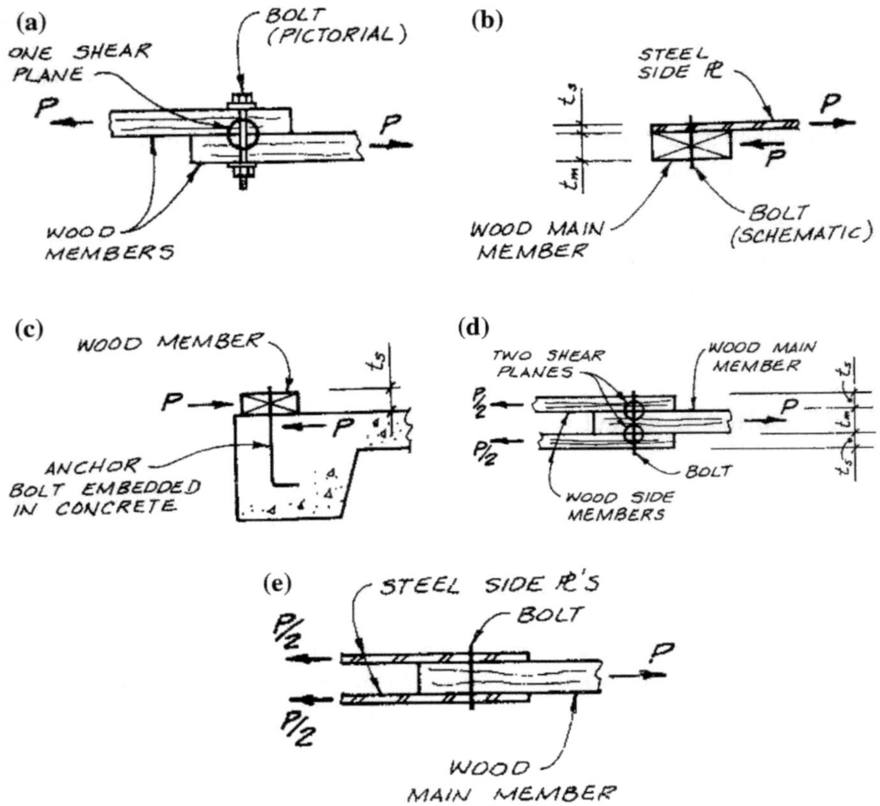

Fig. 11.11 **a** Wood-to-wood single-shear bolted connection. **b** Wood-to-metal single-shear bolted connection. **c** Wood-to-concrete anchor-bolted connection. **d** Double-shear wood-to-wood bolted connection. **e** Double-shear wood-to-metal bolted connection

Bolt sizes used in wood connections typically range from 12.5 to 25 mm in diameter increments of 3 mm. In the past, large bolts (31 and 37.5 mm in diameter) were also used. However, the problems associated with large-diameter bolts include the difficulty in accurately cutting larger holes and in alighting large-diameter bolts in a multiple-bolt connection.

In shear connections, the load is perpendicular to the axis of the bolt. The angle of load to grain in the wood can be zero (load parallel to grain), 90° (load perpendicular to grain) or at some intermediate angle Θ. The angle of load to grain affects the design value of bolted connections.

The following points should be noted about bolted connections:

- To create a bolted connection, the faces of two or more members must overlap sufficiently to accommodate the bolt(s) and all required edge and end distances.
- To install a bolt, a hole must be drilled through each of the members it will join. The hole must not be too larger too small or too rough.

- If the hole is too large, the bolt will rotate along its axis, loading the wood non-uniformly, resulting in localized crushing and consequent reduction in the capacity of the bolt.
- If the hole is too small, the wood member may split during installation of the bolt or after shrinkage of the wood around it.
- If the hole is too rough, the area of wood in contact with the bolt will be reduced resulting in localized crushing, resulting eventually in an oversized hole that could prevent the joint from behaving as predicted.
- If the threaded portion of the bolt bears on wood, the effective bearing area will be reduced with the same consequence described for rough bolt holes.
- Washers of adequate size should be provided between wood member and the bolt head, and between the wood member and the nut. Washers can be circular or square. The size of the washer is not particularly critical for bolts loaded in shear. The washer simply protects the wood as the nut is tightened, i.e. it prevents localized crushing. The nut should be snuggled tight, but not too tight, i.e. the washer should not be embedded in the wood member by over tightening the nut.
- It is generally necessary to have holes a minimum of 0.8 mm larger than the bolt diameter, to avoid development of shrinkage stresses in the wood around the bolt. To accommodate the shrinkage that may occur after a building is enclosed, and during the first heating season (in the temperate countries), it is recommended that all loose bolts be tightened one year after installation.

When bolted joints are loaded at an angle to the grain as shown in Fig. 11.12, the safe compression strength parallel to the grain is obtained using the Hankinson formula earlier stated in Chap. 3. When bolts are loaded perpendicular to their axis, they bear upon the sides of the bolt hole, referred to as the projected area of the bolt. If the bolts are large and stiff, this bearing pressure is uniform, and the allowable bolt load can be computed as the product of allowable bearing pressure (typically 8274 kPa for wood) and the projected bolt area [length of the bolt in main member (i.e. penetration) x bolt diameter]. If the bolt is slender, it will not be stiff enough to distribute the bearing load uniformly, and the bearing stress will concentrate in that portion of the bearing surface near the surfaces of the member.

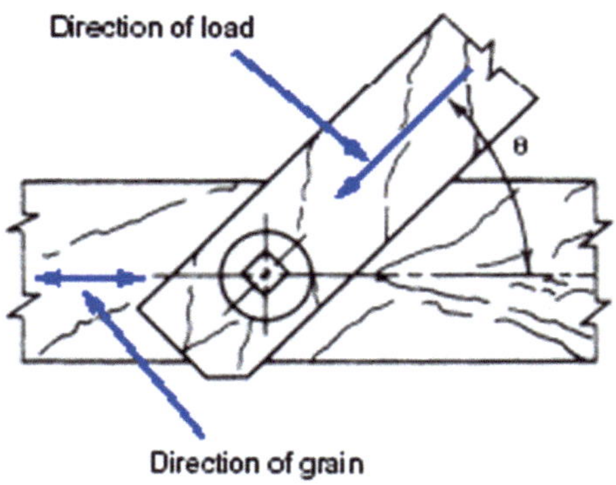

Fig. 11.12 Bolted joint loaded at an angle to the grain

11.8 Lag Bolts

Lag bolts are relatively large-diameter fasteners that have a wood screw thread and a square or hexagonal bolt head. The typical lag bolt (Fig. 11.13a) is 9.375 mm or greater in diameter. Such bolts are installed in pre-bored holes that accommodate the shank and the threaded portion as shown in Fig. 10.13b. The larger diameter hole should be of the same diameter and length as the unthreaded shank of the lag bolt while the diameter of the lead hole should be a percentage of the shank diameter, depending on the specific gravity of the wood. Lag bolts are used in both shear and withdrawal connections and also in either wood–wood or wood–metal connections as shown in Fig. 11.14a and b when an excessively long bolt would be required to make a connection or when one side of the bolted connection may be inaccessible.

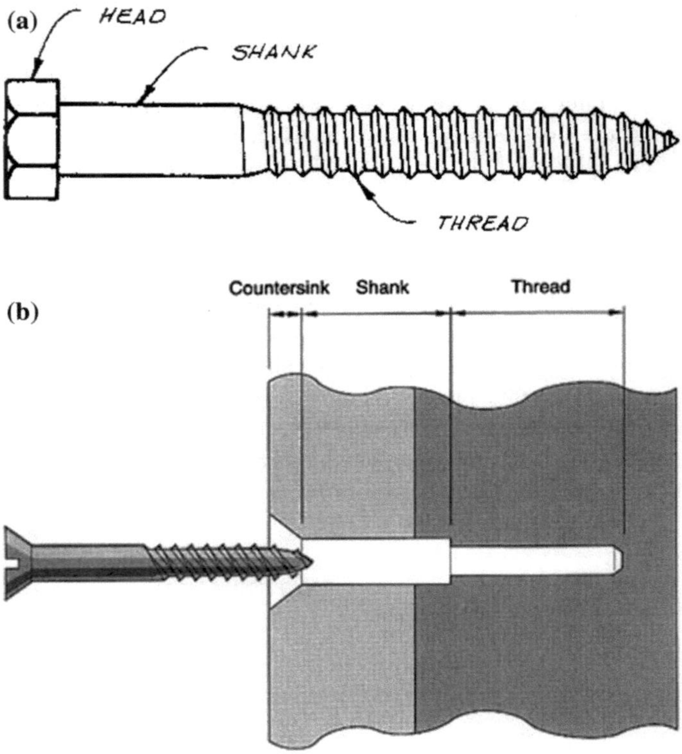

Fig. 11.13 **a** A typical lag bolt. **b** A lag bolt installation

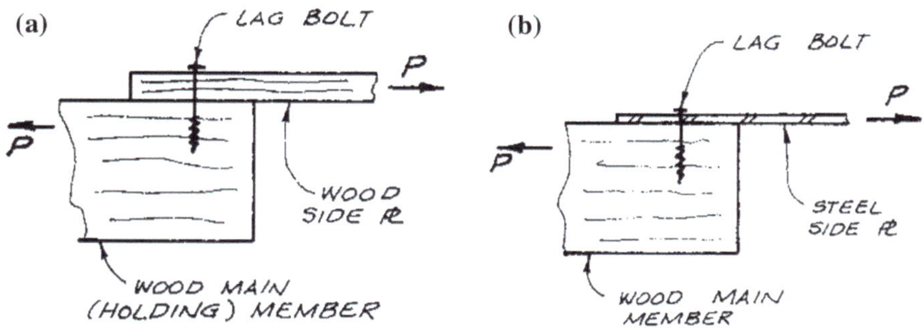

Fig. 11.14 **a** Wood-to-wood lag bolt connection (single shear). **b** Wood-to-metal lag bolt connection (single shear)

11.9 Wood Screws

Wood screws shown in Fig. 11.15 are usually used in fastening millwork where resistance to withdrawal is a requirement. They also find some applications in structural framing. Screws are designed to be much better at resisting withdrawal than nails. However, they are usually more expensive than nails because of the machining required to make the thread and the head. When used for structural purposes, it is better that screws not be loaded in withdrawal.

11.10 Split Ring Connectors

Split rings (Fig. 11.16) are connectors installed in pre-cut grooves in wood members to provide a large bearing surface to resist shearing forces in the wood connection. They are typically manufactured in two sizes, 62.5 and 100 mm, and

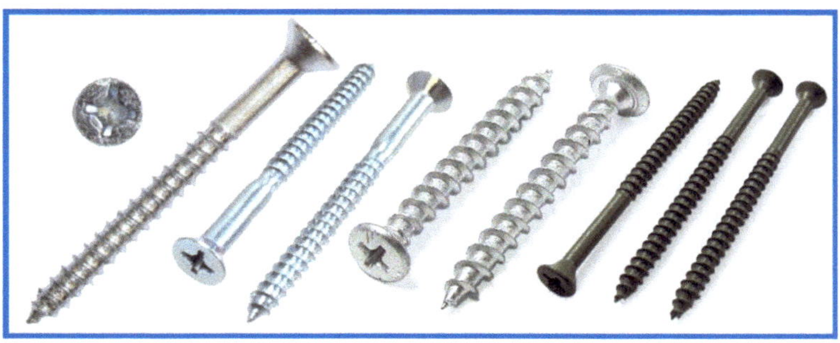

Fig. 11.15 Wood screws

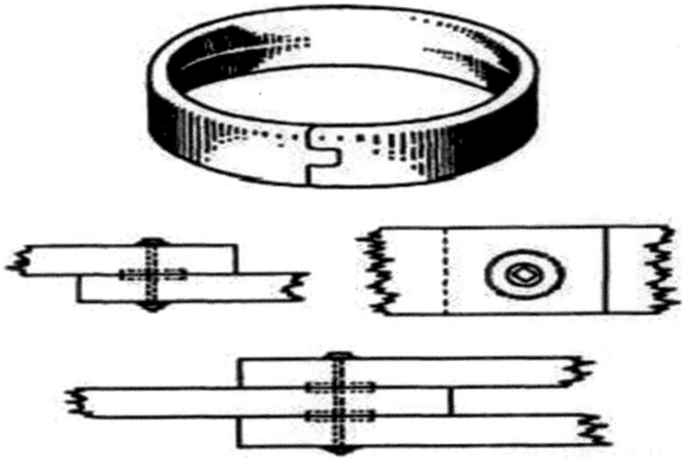

Fig. 11.16 A split ring connector

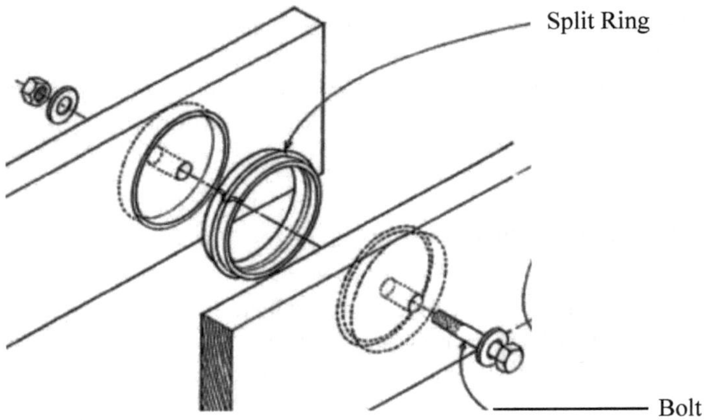

Fig. 11.17 A split ring connection assembly

are meant for only for wood-to-wood connections because the steel ring should fit into a groove cut into the mating surfaces of the wooden members being connected. As shown in Fig. 11.17, a bolt or lag bolt is required at the centre of the split ring merely to hold the assembly together. The bolt itself does not function in bearing.

11.11 Shear Plates

Shear plates (Fig. 11.18) are cup-shaped connectors which fit into pre-bored grooves in the wood members. They are intended for use in members that must be disassembled and reassembled or in making connections between wood members and steel members (Fig. 11.19). They are commonly used to attach steel straps and columns, to construct trusses with steel gusset plates for heel straps in trusses, and for field connections between glulam members of timber structures.

Fig. 11.18 Shear plates

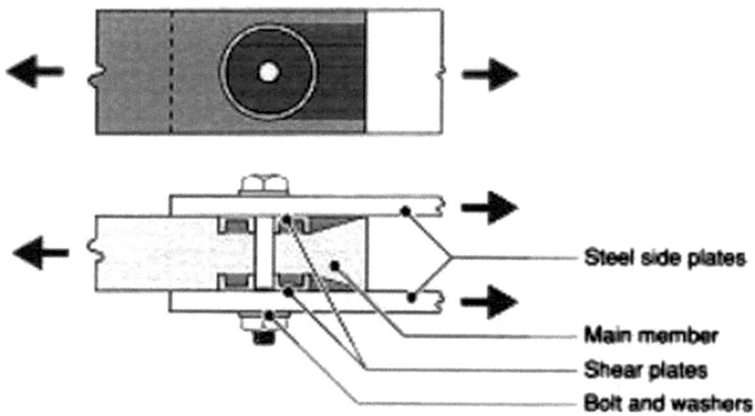

Fig. 11.19 Wood-to-metal shear plate connection (double shear)

11.12 Nailed Gusset Plates

Sometimes, to increase the joint area available for nailing, thus permitting greater load carrying capacity and reducing the possibility of wood-splitting, plywood or metal plates are used. They usually require smaller nails but a greater nailing intensity, at times twice the amount of nails required for a simple nailed joint. In many parts of tropical Africa, plywood gusset plates (Fig. 11.20) are more commonly used than metal plates, particularly in wooden truss fabrication as illustrated in Fig. 11.21 and would be further discussed in the next chapter.

11.13 Framing Anchors

There are numerous types of proprietary wood-framing anchors, commonly used to eliminate the challenges associated with toenailing and withdrawal loads and also to provide stronger shear connections. The anchors are typically produced using light gauge metals, and different manufacturers would specify their respective allowable load values which are typically not codified.

11.14 Choice of Fastening

The selection of a fastener often depends on a number of interrelated factors. Apart from cost considerations, the major structurally related factors include lumber thickness, method of fabrication, assembly or erection problems, type of loading

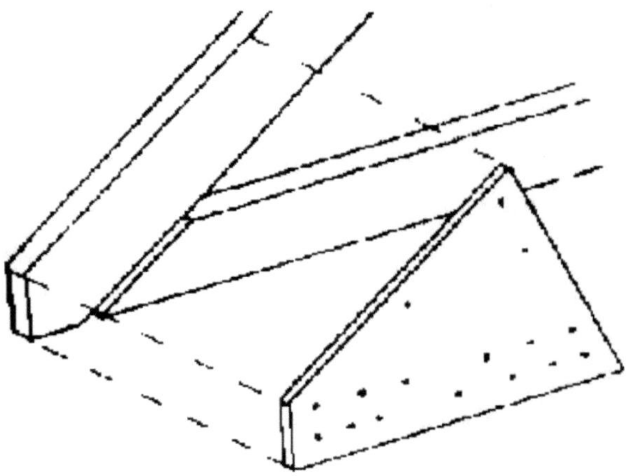

Fig. 11.20 A nailed plywood gusset plate connection

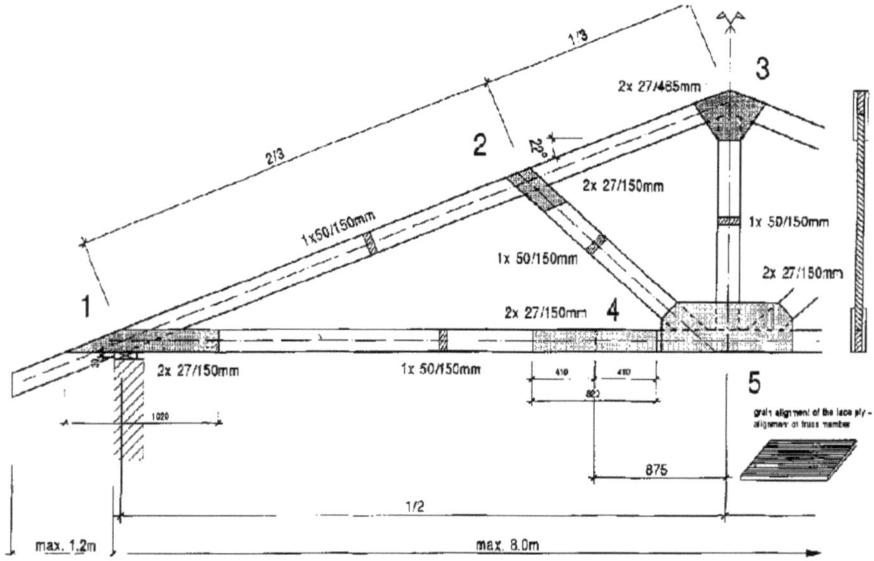

Fig. 11.21 Details of a wooden truss jointed with plywood gusset plates

and the load-bearing capacity required. For the 50 mm and greater lumber thicknesses most frequently encountered in structural work, nearly all of the common fastenings are suitable. Nails and spikes are seldom used in framed assemblies of lumber that are thicker than 50 mm because of the wood-splitting problems commonly encountered with the use of the relatively large and long nails required in such situations. It should be noted, however, that the tendency of a wood member to split, though a common phenomenon encountered in the nailing of tropical hardwoods, varies with the species and moisture content, so that no precise nailing rules are possible. Also, in general, the fewer the number of fasteners required, the better. For example, the use of a small number of large bolts is preferable to the use of many small bolts from the standpoint of labour requirement and fabrication accuracy. The method of fabrication itself may again favour one fastener over others. For example, if the fabrication process is such that members have to be laid together before the holes are bored, bolts have an advantage because the disassembly that would be mandatory if other connectors were to be used would not be necessary.

The type of loading comes in as a selection factor in so far as it affects increase in allowable loads for the duration of loading, the effect of which varies for different types of fasteners. Vibratory loads which tend to loosen some fastenings are also a critical selection factor. As regards load-bearing capacity, nails, wood screws and connector plates are generally employed for light-frame construction, while bolts, split rings, and other connectors are largely employed in heavy timber construction.

11.15 Timber Joinery

Timber joinery is a traditional method of connecting wood members without the use of metal fasteners. Although the use of metal fasteners for connections is almost universal, timber joinery still offers a unique visual appearance exhibiting a high degree of craftsmanship. While metal fasteners permit the use of moderate-sized members to carry and transfer loads, their installation requires only minimal removal of wood fibre, timber joinery and requires the removal of significant wood fibre where joints occur. For this reason, the adequacy of timber members is usually governed by the connections. Increased member size in relation to what would be required for construction employing metal fasters is often required. In addition, wood engineering codes do not provide specific load transfer information for timber joinery due to sensitivity to quality of workmanship and material quality. Also, the amount of skill and time required for measuring, fitting, cutting and trial assembly is far greater for timber joinery than for other types of wood construction.

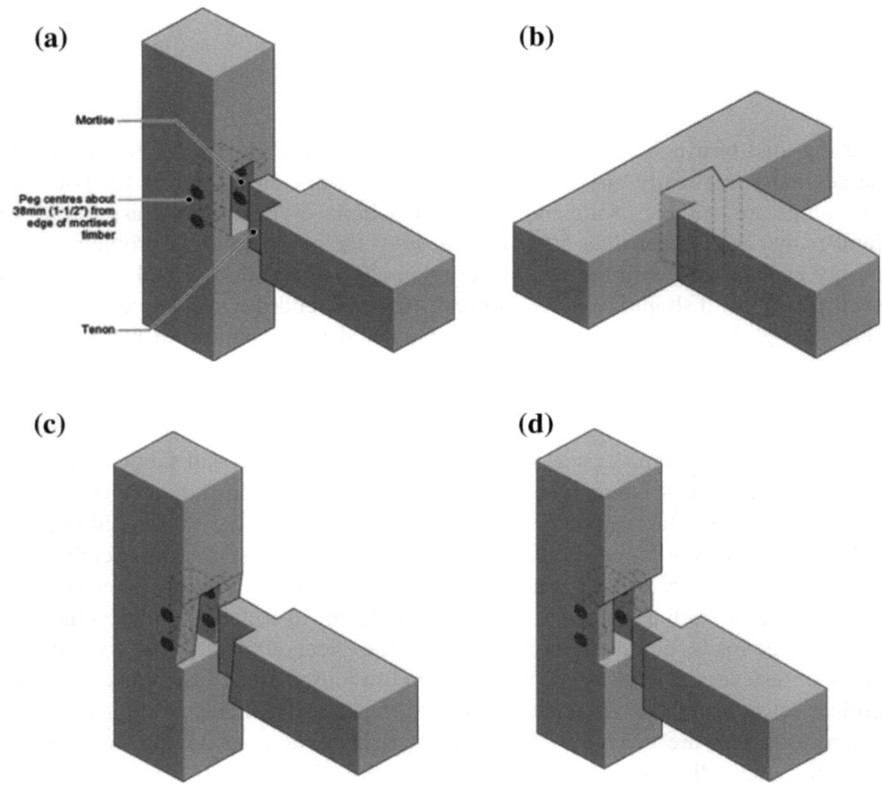

Fig. 11.22 **a** Basic mortise and tenon joint. **b** Dovetail joint. **c** Bevelled shoulder joint. **d** Housed mortise and tenon joint

Therefore, it is not the most economical means of connecting the members of wood buildings. Figure 11.22a–d shows some of the basic joints for timber joinery construction.

11.16 Wood Adhesives

Adhesives play a prominent role in wood construction. They are used for the manufacture of laminated products; as a means of increasing the structural rigidity of sheathing/joist combinations in floors and of affixing non-structural panel products; end joining dimension lumber; and repair. Adhesives made from synthetic resins produce the most efficient form of joint, as strong as or even stronger than the timber joined, and many are immune to attack by dampness and decay. With this type of joint, all contact surfaces must be planed smooth, and the necessary pressure provided during setting of the glue. The members may be glued directly to each other using lapped joints, or single-thickness construction may be used by the adoption of gussets. As with nailed joints, lapped members may not provide sufficient gluing area, and gussets must then be used to provide the extra area.

Structural composites such as plywood, oriented OSB and wafer-board, prefabricated wood I-joists, PSL, LVL and glulam are dependent upon adhesives to transfer the stresses between adjoining wood fibres. Interior-use wood products such as particleboard, which is used for furniture and for some structural applications such as flooring underlay, and hardwood plywood, which is used for furniture and decorative panelling, also rely on adhesives for laminating wood material. The selection, application rate and curing conditions for adhesives for these products are controlled at the point of manufacture. A brief discussion of the principal adhesives used in these products is presented to address questions which sometimes arise about permanence of bond, reliability, resistance to environmental factors and emission of volatile chemicals into buildings.

There are two principle types of adhesive used for the manufacture of wood products. They are urea-formaldehyde (UF) which is suitable for interior-use products and phenol-formaldehyde (PF) which is used for exterior applications. UF adhesive is thick, creamy syrup which cures to a colourless solid. UF adhesives are very economical and fast curing but are not suitable for damp conditions. For this reason, they are used for panels intended for non-structural use such as particleboard and hardwood plywood. UF adhesives are non-staining and therefore have the further advantage of not blemishing the high-quality expensive face veneers used for hardwood panels for interior finish applications. The raw materials for UF adhesives are derived from natural gas through the intermediates of ammonia for urea and methanol for formaldehyde.

PF adhesives are dark purple-brown in colour and give the dark glue lines associated with products such as plywood and OSB. Known as the phenolics, they are a derivative of crude oil and the principal resins approved for the manufacture of wood products intended for exterior applications. They are also are used for the manufacture of glulam, PSL, LVL, plywood and OSB/wafer-board. Formaldehyde is an allergic irritant to some people when the time and level of exposure are high. Effects are compounded when a building has air change rates below accepted standards. PF adhesives are also somewhat more expensive than UF adhesives and exhibit lower levels of formaldehyde emissions. However, the level of formaldehyde emission from any new product is time dependent. Emission level is highest when the product is new and decreases steadily as the time in service increases. Various types of extenders such as walnut shell flour, bark flour and wood flour are used to moderate the cost of PF glues, control penetration into the wood fibre and moderate strength properties to suit the materials being bonded. Resorcinol–formaldehyde (RF) adhesive is a phenolic substance which is more reactive than the PF adhesives. Being more reactive, curing is faster and takes place at room temperature and below. Otherwise, these glues have the same basic properties as the PF adhesives. However, high cost of the resorcinols means in practice that they are often blended with the PF adhesives to moderate the cost.

There are many adhesives available to improve the structural performance of building elements or to apply non-structural panels in a way that does not leave surface blemish on the panel, as would nailing. These field-applied tube-type adhesives are available in many types suitable for both interior use and exterior use. The recommended application temperature can range between −10 and 40 °C. These fast-setting adhesives can be used to bond wood and panels to metal, gypsum wallboard, concrete and foam insulation. Where floor sheathing is affixed to joists with a field-applied elastomeric adhesive in addition to nails or screws, improved vibration and deflection can be gained. In this situation, the joists and sheathing act as a single composite section, and increases in spans of 5–10% when compared to nailed or screwed floors are possible.

As with all building materials, wood is subject to damage if exposed for a long period of time to adverse conditions, if original design was faulty, or if overstressed due to loadings beyond design specifications. Damage assessment requires the analysis of a structural engineer. In some cases, on-site repair rather than replacement may be possible. Where a repair is recommended by a structural engineer qualified in designing and supervising repairs, it may require the use of epoxy adhesives. The procedure for such repair will usually mean sealing the area to be repaired using a high viscosity epoxy with putty-like properties which is trowelled over cracks and holes to contain the epoxy repair material. The putty is also used to embed injection and vent port hoses to accept pressure injection equipment.

Worked Examples

1. A rope line for hanging clothes is to be erected using 4 No. 75 × 125 mm Mansonia wooden columns. To each column, a short piece of 25 × 50 mm teak wood is to be fastened with common wire nails. If, based on the tensile strength of the rope, the maximum weight of wet clothes to be hung on the line at any time is 1500 kg, determine the number of nails required at each joint to prevent

shear failure. Assume that a 4.06-mm-diameter nail is selected, the lateral load per nail is 628 N, and the total load is to be equally shared by each of the four joints.

Solution

Load per joint = (1500 × 10) N/4 = 3750 N
No of nails required per joint = 3750/628 = 6 nails.

2. Determine the number of 19-mm-diameter bolts required to fasten a 75-mm Mansonia piece to another 100-mm piece subjected to a lateral load of 7.5 kN. Assume that the tabulated design value for the basic load parallel to grain on one 19-mm bolt is 5449 N.

Solution

Thickness of thinner member = 75 mm
Bolt diameter = 19 mm
No of bolts = 7500 N/5449 N = 6.4 = 7 bolts.

3. A common wire nail is used to fabricate the joint sketched below. The 25 × 50 mm side member and 50 × 150 mm main member were both fabricated from *Milicea excelsa*. Assuming that the length (L) and diameter (D) of the common wire nail are 75 and 3.7 mm, respectively; the safe lateral load is 523 N per nail, while the minimum penetration requirement for connection is 10D, determine the adequacy of the penetration of the nail in the main member and the number of such nails required to withstand a shear force of 5 kN.

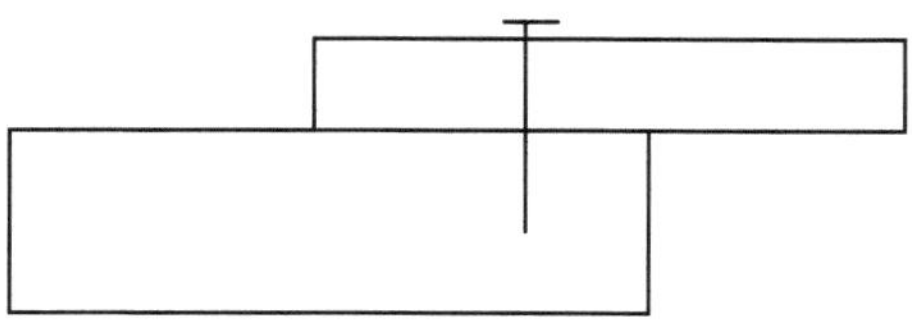

Solution

- Compute the expected minimum penetration for group II species, i.e. p = 10D where D = nail diameter

The minimum penetration requirement = 10 × 3.7 = 37 mm

- Compare 10D with $p = l - t_s$. If $p > 10D$, nail is ok

  Actual penetration, $P = L - t_s = 75 - 5 = 50$ mm
  It means that the nail is OK for the connection

- Divide expected load by the safe lateral load to determine number of nails required

  No of nails required = 15000/523 = 9.6, say 10 nails

4. A 12.5-mm-diameter bolt is used to fasten two wood members together. If the penetration of the bolt in the mail member is 50 mm, compute the allowable bolt load. Assume that the wood-bearing strength is 8274 kPa.

Solution

$$\begin{aligned}\text{Projected bolt area} &= \text{bolt penetration} \times \text{bolt diameter} = 50 \times 12.5 = 625\,\text{mm}^2 = 0.0625\,\text{m}^2 \\ \text{Allowable bolt load} &= \text{wood bearing strength} \times \text{projected area} = 8274 \times 0.0625 \\ &= 517.13\,\text{kPa}.\end{aligned}$$

Practice Questions

1. A common wire nail is used to fabricate the joint between the 50 × 150 mm *Afzelia africana* side member and 150 × 150 mm main member sketched below, where both fabricated from *Milicea excelsa*. Assuming that the length (L) and diameter (D) of the common wire nail are 125 and 5.625 mm, respectively, the safe lateral load is 980 N per nail, while the minimum penetration requirement for connection is 10D, determine the adequacy of the penetration of the nail in the main member and the number of such nails required to withstand a shear force of 10 kN.

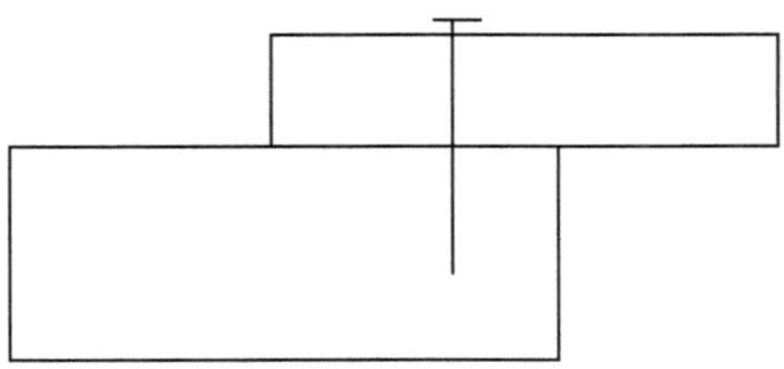

2. Determine the number of box nails required to carry a lateral load of 15 kN between 50-mm and 100-mm-thick pieces of wood. Assume that the length (L) and diameter (D) of the box nail are 112.5 and 4.8 mm, respectively, the minimum penetration requirement is 11D, while the safe lateral load per nail is 775 N.

3. If a 50 × 100 mm wooden member is toe-nailed using a 125-mm-long box nail to another 75 × 100 mm wooden member, determine the penetration of the toe nail in the main member if (a) the connection is laterally loaded and (b) the connection is loaded in withdrawal.
4. A number of 20d nails are to be used to connect a *Terminalia superba* piece, 25 mm thick, to another piece, 100 mm thick. Determine the penetration and the number of such nails if the connection is to be subjected to: (i) a pull of 15 kN along the length of the nail; (ii) a force of 15 kN along the grain of the planks. Assume that for 20d nails, length = 101.6 mm, diameter = 4.88 mm, safe withdrawal load per millimetre of penetration = 17 N/mm, safe lateral load per nail = 511 N and minimum penetration required for safe lateral loading = 13D.
5. Determine the number of 137.5-mm-long hardened nails required to carry a lateral load of 25 kN between 75-mm and 300-mm-thick pieces of *Terminalia ivorensis,* and the number of such nails required if the connection is to be subjected to pull along the length of the nails. Assume that nail diameter = 4.425 mm, safe withdrawal load per millimetre of penetration = 25 N/mm, safe lateral load per nail = 600 N and minimum penetration required for safe lateral loading = 13D.
6. A 1.56-mm-thick steel gusset plate used to splice two pieces of wood together is fastened to the wood with 16 No. common nails to their full penetration. If the lateral load capability of each nail is 265 N, what is the maximum load the joint can carry?
7. A 25-mm-diameter bolt is used to fasten two wood members together. If the penetration of the bolt in the mail member is 100 mm, compute the allowable bolt load. Assume that the wood-bearing strength is 8274 kPa.

References

Breyer DE, Fridley KJ, Cobeen KE (1999) Design of wood structures, 4th edn. McGraw-Hill, New York, USA

Faherty KF, Williamson TG (1999) Wood engineering and construction handbook, 3rd edn. McGraw-Hill Handbooks, New York, USA

Goldstein EW (1999) Timber construction for architects and builders. McGraw-Hill Publishers, New York, USA

Gurfinkel G (1973) Wood engineering. Published by Southern Forest Products Association, USA

Khurmi RS, Khurmi N (2013) Strength of materials (Mechanics of solids in S.I. units). International edition. S. Chand & Company PVT Ltd., New Delhi, India

Lindley JA, Whitaker JH (1996) Agricultural buildings and structures, revised edition. Published by the American Society of Agricultural Engineers, USA

Marshall WT, Nelson HM (1978) Structures. Published by Pitman Publishing Ltd., London

Midwest Plan Service (1977) Structures and environment handbook. Published by Midwest Plan Service, Ames, Iowa, USA

NCP (2005) Nigerian code of practice for structural timber design. Published by the Standards Organisation of Nigeria

Ozelton EC, Baird JA (1976) Timber designers' manual. Granada Publishing Limited, London

Seward D (2003) Understanding structures, analysis, materials, design, 3rd edn. Palgrave Macmillan, New York, USA

Shaeffer RE (1999) Elementary structures for architects and builders, 4th edn. Prentice-Hall Publishing Co., New Jersey, USA

Singh S (1982) Strength of materials, 2nd revised edition. Vikas Publishing House PVT Ltd., Noida, U.P., India

Willenbrock JH, Manbeck HB, Suchar MG (1998) Residential building design and construction. Prentice Hall, New Jersey, USA

Chapter 12
Design of Wooden Trusses

A truss is a structural frame, composed of members, commonly referred to as webs and chords, which are assembled to form one or more connected triangles, thus producing a rigid frame. It relies on the triangular arrangement of its members to transfer loads to reaction points. Roof trusses consist of sloping rafters which meet at the ridge, a main tie connecting the feet of the rafters, and internal bracing members. Both pitched and flat roof configurations are possible as shown in Fig. 12.1a, b. They are used to support a roof covering in conjunction with purling. Members are laid longitudinally across the rafters, the roof covering being attached to the purling. The arrangement of internal bracing depends on the span. Rafters (shown in Fig. 12.2) are normally divided into equal lengths, and ideally, the purlins are supported at the joints, so that the rafters are only subjected to axial forces. This is not always practicable, since purlin spacing is dependent on the type of roof covering.

Wooden trusses are now regarded as the most economical way of providing pitched roofs of spans up to 12 m. The most economic depth for a truss is usually around 1/6th to 1/8th of the span, while angles between adjacent members should be kept between 30° and 60° where possible. Several styles of trusses have evolved over time as will be discussed shortly. The usual aim is to minimize the length of the compression members (top chord) because the strength of a strut is inversely proportional to the square of its length, but with tension members or the ties (bottom chord), the length is irrelevant to the load-carrying capacity. The internal bracing members are triangulated and, as much as possible, arranged so that long members are in tension and compression members are short to avoid buckling problems.

A.O. Olorunnisola, *Design of Structural Elements with Tropical Hardwoods*,
DOI 10.1007/978-3-319-65343-3_12

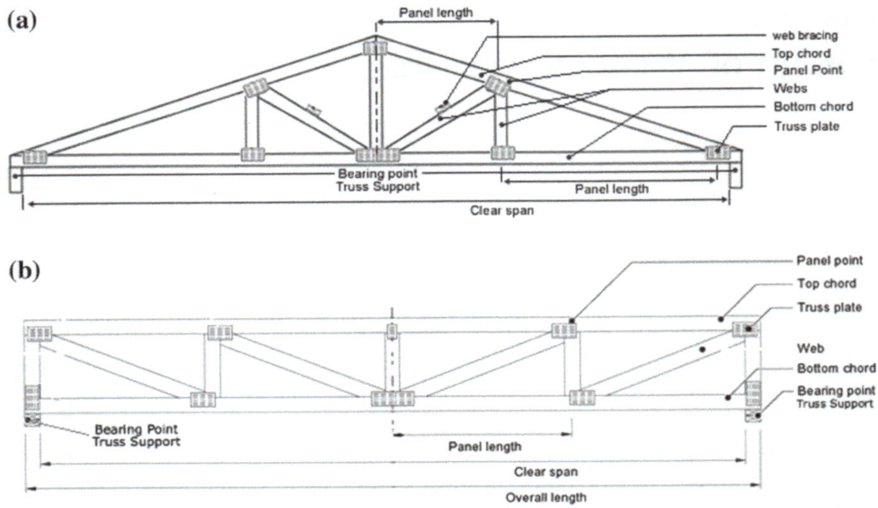

Fig. 12.1 **a** Schematic diagram of a pitched (*triangular*) wooden truss. **b** Schematic diagram of a parallel chord (*flat*) wooden truss

Fig. 12.2 A wooden truss installation in progress

12.1 Advantages of Wooden Trusses

Some of the advantages of using trusses include the following:

- **Strength**: Trusses are, in general, capable of supporting heavy load over a considerable span. The triangular arrangement of the members gives trusses high strength-to-weight ratios, which permit longer spans than conventional framing, and offers greater flexibility in floor plan layouts. They do not sway or deflect as much as solid structures of equivalent weight.

- **Economy**: Through efficient use of wood and by providing a system that is installed in as little as half the time of conventional wood framing, wooden trusses provide an economical framing solution. They are an efficient way of minimizing material weight, and they provide a strong and efficient structural wood system specifically engineered for each application.
- **Environmental benefits**: Wooden trusses enhance wood's environmental advantages by optimizing wood use for each specific application.
- **Versatility**: Wooden trusses can be designed in almost any shape or size, restricted only by manufacturing capabilities, shipping limitations and handling considerations. Complex shapes and unusual designs are easily accommodated. They are also well suited to most building construction, i.e. single- and multi-family residential, institutional, agricultural and commercial construction. Their long span capabilities can offer designers the flexibility of floor layouts without interior supports. Besides, wooden trusses are compatible with other structural products. They can be connected to other trusses (i.e. girder trusses) or combined with other components, such as glulam, LVL and PSL. The use of roundwood timber as well as poles and lumber from fruit trees for roofing is very common across tropical Africa. A study conducted at the University of Ibadan, Nigeria, showed that wood from the branches of guava tree, a medium density wood that is very resistant to termite, insect and fungal attack, is a suitable and relatively cheaper material for truss fabrication. The obvious limitations that could be associated with the use of such round timber for truss fabrication, i.e. relatively heavy weight, twisted or curved shapes, can be overcome through careful selection of suitable species and proper arrangement of members. The bowed shape of the branches which may be considered one of the limitations of poles derived from the tree branches could be of great advantage in cambering. A sample of the truss fabricated from guava tree is shown in Fig. 12.3.

Fig. 12.3 Side view of the truss fabricated with branches of guava tree. *Source* Lucas et al. (2006)

- **Ease of connection**: Wooden roof trusses are commonly supported on concrete or masonry walls using simply installed connections to join the roof to the walls.
- **Easy installation of building services**: The open web configuration of roof and floor trusses allows easy placement of plumbing, electrical, mechanical and sanitary services.
- **Enhanced architectural and structural innovation**: Considerable architectural and structural innovation is possible due to the availability of a wide range of trusses and connections. Architectural trusses, or exposed trusses with attractive detailing, can be used effectively as a feature to suit the architectural theme of the building. They can convey an impression of strength and tradition or a feeling of lightness and space. They can be left natural, painted, oiled, stained or highly decorated.
- **Prefabrication advantages**: Most wooden trusses are pre-fabricated to standard designs using pressed metal nail plates, providing versatility in design, economy, fast delivery and ease of erection. Hinged connector plates used with mono-pitch trusses allow modular homes to be assembled with conventional roof pitches, greatly enhancing their appearance.

12.2 Disadvantages of Wooden Trusses

Some of the disadvantages of wooden trusses include the following:

- They tend to take up more room than solid structures of the same strength.
- Only a few carpenters can frame any wooden truss for anything more complicated than the simplest roofs.
- The house must be built precisely as designed; otherwise, the trusses do not fit.
- Because of the complexities involved in building the roof structures with multipronged fasteners with the help of hydraulic equipment, these structures cannot be built at the site of the building. Hence, these roofs are designed at a plant and have to be transported to a location of the building. Transportation itself raises a major drawback. Once the trusses are assembled, they can get a little too big. Sometimes, they are too big for a truck. In such cases, specially designed truss trailers have to be used to haul the structures around.
- The conventional truss design leaves a large volume of attic space unused as these structures create a web of wood framing that leaves no possible space of using it as a utility. Further, because of the entangled wooden framework, the aesthetic appearance of the roof is very low. No doubt, it can be covered up with wooden trim, but it is an additional cost.
- Lastly, wooden trusses are susceptible to fire. Light gauge steel frames, which constitute an alternative, are more resistant and more robust when compared to wooden roof trusses.

The important points to take care of during the construction of the roof include its peak, clear span, pitch, the overall length and the overhang. Ensure that the lumber is not too moist and is adequately flame-retardant, and the metal connectors used are rust-free.

12.3 Truss Shapes

Wooden trusses can be fabricated in a nearly infinite number of sizes and shapes. Truss configurations are distinguished by the orientation of their diagonal web members, but more importantly by the senses of the forces in all their web members. Several of the more common configurations bear the names of their developers such as Howe and Pratt. Variations intended for floor construction are known as flat Howe and Flat Pratt Trusses. Schematic diagrams of these commonly used trusses are shown in Fig. 12.4a–l. The different types are suitable for different purposes. For example, the Fink or 'W'-type truss is suitable for spans of up to about 12 m. The King Post truss is simple and economical for relatively short spans; Pratt or flat truss is used in roofs or floors. It may be designed as top or bottom chord bearing, or for simple or multiple spans. It may also be cantilevered at one or both ends. They may be ordered with a built shallow slope to offset deflection and to provide positive drainage when used as a flat roof system. The mono-shape may be simple span, multiple span or cantilevered, and top chord bearing is possible. The inverted truss is used to provide a vaulted ceiling along a portion of the span. The sloping chord flat truss, which is also referred to as a high heel common truss, is used to provide positive drainage to both sides of the building. The hip truss, also referred to as a step-up hip truss, is used to create hip roofs. The scissor truss is used to create a vaulted ceiling along the entire span; the slope of the bottom chord is usually equal to 1/2 of the slope of the top chord. Large scissor trusses are often shipped in two pieces and field spliced. Scissors truss is recommended when a high ceiling clearance is required. The gambrel truss is used to create a gambrel or barn-shaped roof profile.

12.4 Truss Plates

Nowadays, in many developed countries, wooden trusses are connected by means of galvanized steel plates referred to commonly as truss plates or connector plates (Fig. 12.5). The plates are manufactured by high-speed stamping machines that punch out the plate teeth and shear the plate to required size. Many sizes and gauges of connector plates are manufactured to suit a variety of joint geometries and loadings. The use of metal plates permits the plant fabrication of trusses with consistent and dependable engineering properties.

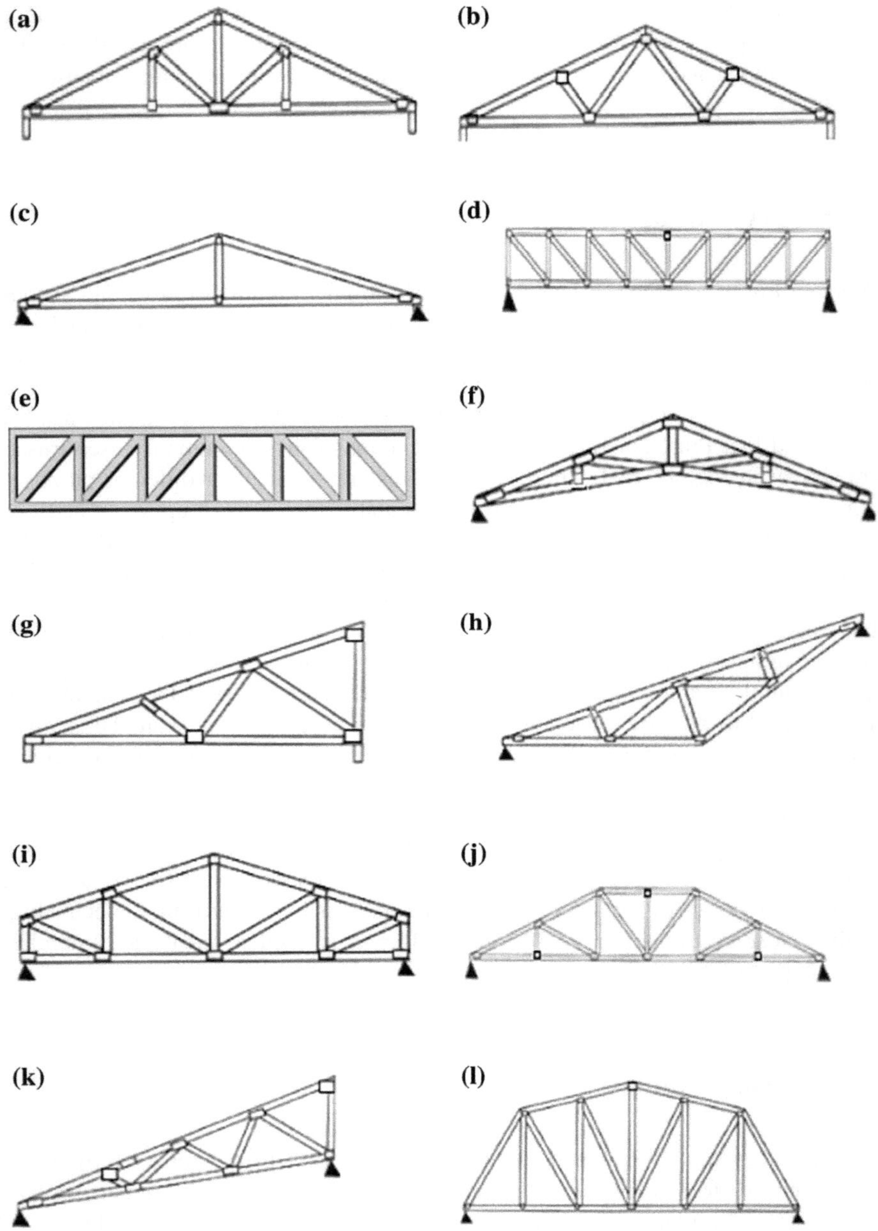

Fig. 12.4 **a** Howe truss. **b** Fink truss. **c** Kingpost truss. **d** Flat pratt truss. **e** Flat howe truss. **f** Howe truss. **g** Mono truss. **h** Inverted truss. **i** Sloping chord flat truss. **j** Hip truss. **k** Mono scissors truss. **l** Gambrel truss

Fig. 12.5 Truss metal plates

The metal connector plate transfers loads between adjoining members through the connector plate teeth. The connector plate strength is dependent on the grip of the teeth and the shear and tensile capacity of the steel plate. Each plate must be installed using specifically designed press or roller truss plate equipment to achieve published design values. Plate widths can be from 25 to 300 mm, and lengths can be up to 600 mm or even longer. Teeth lengths vary from about 6–25 mm. Nail-on plates are occasionally provided to allow assembly by the builder on the site. For example, nail-on plates are sometimes used to join separate parts of a field-assembled truss.

12.5 Truss Spacing and Jointing

Unless there are particular constructional requirements, roof trusses should, as much as possible, be spaced to achieve a minimum of weight and economy of materials used in the total roof structure. As the distance between trusses is increased, the weight of the purlins tends to increase more rapidly than that of the trusses. For spans up to about 20 m, the spacing of wooden trusses (shown in Fig. 12.6) is likely to be about 2 m. The pitch, or slope, of a roof depends on locality, imposed loading and type of covering. Heavy rainfall may require steep slopes for rapid drainage; a slope of 22° is common for corrugated steel roofing sheets. Manufacturers of roofing material usually make recommendations regarding suitable slopes and fixings. Jointing is usually done using any of the following:

- Bolted connections to transfer load directly from member to member. They work best where some of the chords are two-element members to allow the joining element to be accommodated between the two pieces. Bolted trusses in unseasoned hardwood can provide low-cost solutions for smaller spans where economy is the major consideration.

Fig. 12.6 Wooden truss spacing

- Plywood gussets can be used to join elements using nails.
- Steel nail plates provide a cost-effective means of pre-fabricating trusses but are less suited where the trusses must be manufactured on site.
- Very large truss elements may be joined using fin-plate connections or glued-in dowels. In both these cases, it is wise to build and test prototypes of the connections to ensure that adequate strength can be achieved in practice.
- Glulam timber trusses with bolted plate or split ring connectors can be used as an architectural feature.

12.6 Computation of Joint Loads for a Roof Truss

In calculating joint loads for roof trusses, it usually is assumed that the purlins (shown in Fig. 12.7) transmit the roof load to the truss at the joints. The key factors that determine load distribution are the centre-to-centre spacing of the trusses and the length of the rafter, both of which determine the tributary load per joint. The following examples illustrate the methods of obtaining joint loads.

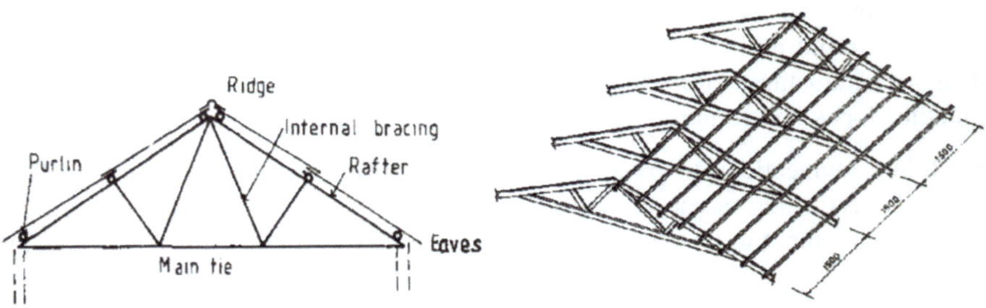

Fig. 12.7 A typical truss geometry and spacing

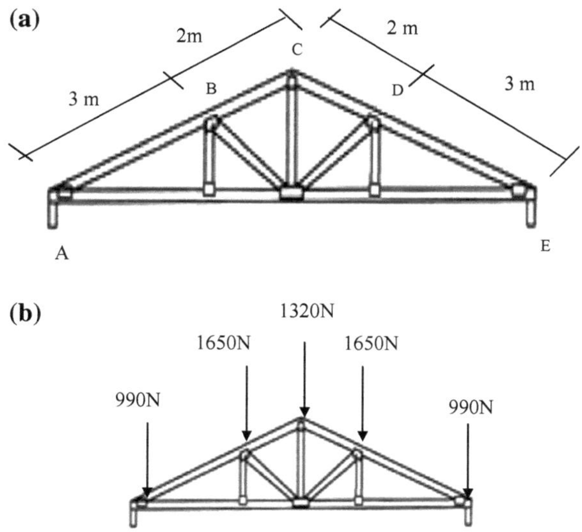

Fig. 12.8 **a** Howe truss. **b** Joint load distribution on the howe truss

Example 1 Calculate the joint loads for each of a Howe roof truss shown in Fig. 12.8a. Assume that the roof covering is composed of the materials indicated in the list of weights, the total length of each rafter (the diagonal members of each truss) is 5 m, and the trusses are to be spaced 1.5 m on centre. The self-weight of the truss is usually added into the roof-covering load, and for this, a provisional estimate must be made. Experience of previous similar calculations will indicate the allowance to be made. For this particular example, let us assume a self-weight of 2400 N.

Loading

$$\begin{aligned}\text{Purlins and ridge} &= 50\text{N/m}^2 \text{ of roof area}\\ \text{Common rafters} &= 80\text{N/ m}^2 \text{ of roof area}\\ \text{Roof boarding} &= 150\text{N/m}^2 \text{ of roof area}\\ &\ \ \underline{280\,\text{N/m}^2} \text{ of roof area}\end{aligned}$$

Solution

Since the truss has two rafters of 5 m length and the trusses are to be spaced 1.5 m on centre, then each truss has to support a roof area of (5 + 5) m × 1.5 m = 15 m^2

$$\text{Therefore total roof load per truss} = 15\text{m}^2 \times 280\,\text{N/m}^2 = 4200\,\text{N}$$

$$\text{Total load carried by one truss, including self-weight} = (4200 + 2400)\,\text{N} = \underline{6600\,\text{N}}$$

We now have to divide up this load proportionately between the joints.

Tributary Area of roof for Joint A = 1.5 × 1.5 = 2.25 m^2
Area of roof for Joint B = (1.5 + 1.0) × 1.5 = 3.75 m^2
Area of roof for Joint C = (1.0 + 1.0) × 1.5 = 3.0 m^2
Area of roof for Joint D = (as in Joint B) = 3.75 m^2
Area of roof for joint "E" = (as in Joint A) = 2.25 m^2
15.00 m^2

The load at any particular joint is proportional to the tributary area of roof associated with it.

Therefore,

$$\begin{aligned}\text{Load at Joint A} &= \text{load at Joint E}\\ &= (2.25/15) \times 6600\,\text{N} = 990\,\text{N}\\ \text{Load at Joint B} &= \text{Load at Joint D}\\ &= (3.75/15) \times 6600\,\text{N} = 1650\,\text{N}\\ \text{Load at Joint C} &= (3/15) \times 6600\,\text{N} = 1320\,\text{N}\end{aligned}$$

The joint loads are indicated on the Howe truss in Fig. 12.8b.

CHECK: Total load sustained on the truss = [(2 × 990) + (2 × 1650) + 1320)] N = 6600 N

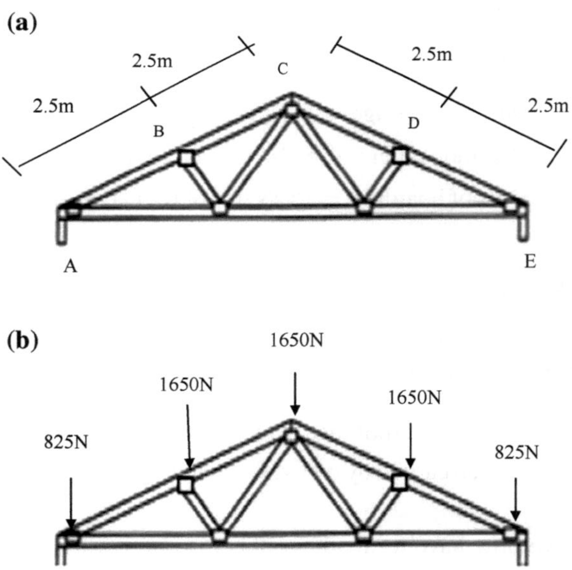

Fig. 12.9 **a** Fink truss. **b** Joint load distribution on the fink truss

Example 2 Assuming the loading and truss spacing data given in Example 1, but a Fink Truss shown in Fig. 12.9a is to be used for the roof design calculate the joint loads for the truss.

Solution

Tributary Area of roof for Joint A = 1.5 × 1.25 = 1.875 m^2
Area of roof for Joint B = (1.25 + 1.25) × 1.5 = 3.75 m^2
Area of roof for Joint C = (1.25 + 1.25) × 1.5 = 3.75 m^2
Area of roof for Joint D = (as in Joint B) = 3.75 m^2
Area of roof for joint "E" = (as in Joint A) = 1.875 m^2
15.00 m^2

The load at any particular joint is proportional to the tributary area of roof associated with it.

Therefore,

$$\begin{aligned} \text{Load at Joint A} &= \text{load at Joint E} \\ &= (1.875/15) \times 6600\,\text{N} = 825\,\text{N} \\ \text{Load at Joint B} &= \text{Load at Joint C} = \text{Load at Joint D} \\ &= (3.75/15) \times 6600\,\text{N} = 1650\,\text{N} \end{aligned}$$

The joint loads are indicated on the Fink truss in Fig. 12.9b.

12.7 Truss Analysis

Truss analysis has to do with determining the magnitude of the tensile and compressive forces in the truss members. We shall be restricting our analysis to plane trusses which means that all of the members and loads lie in one plane. There are perfect and imperfect trusses. A perfect truss is that which is made up of members just sufficient to keep it in equilibrium state when loaded without appreciable change in shape. The simplest form is a triangle which contains three members and three joints. The number of members (bars) of a perfect plane truss can be obtained from the relation:

$$b = 2n - 3$$

where b = number of bars, n = number of joints.

An imperfect truss is that which does not satisfy the condition for a perfect truss. When $b < 2n - 3$, the truss is deficient. When $b > 2n - 3$, such a truss is redundant. There are also issues of determinacy and stability. For a given plane truss of n number of joints, r number of reaction components and b number of bars, the conditions of stability are given as follows:

Condition 1: $b + r = 2n$ for determinate and stable condition.
Condition 2: $b + r < 2n$ for determinate and unstable condition.
Condition 3: $b + r > 2n$ for indeterminate condition.

Forces in members of trusses that satisfy condition 1 are analysed and determined using the principles of statics. It is assumed that the truss is a perfect rigid body; the end points are pin joints; the weight of each member (bar) is negligible except otherwise stated; the truss is loaded only at the joints; and only axial compression and tension loads are applied at the joints also called the node points, although this is not always possible in reality. Also, it is not always possible to establish whether a member is in tension or compression. The direction of the forces may be deduced simply by inspection. If they had been deduced wrongly, it would result in negative numerical values emerging from the calculations. The different methods of truss analysis include algebraic, joints, sectioning, graphical and tension coefficient methods. The details are available for review in textbooks on statics and basic theory of structures. A brief review of the methods of resolution at joints and tension coefficients only are discussed.

Having calculated the reactions and drawn the free body diagram for the truss, the forces in the members may be found, using the method of resolution at joints, by applying the equations of equilibrium joint by joint. Since only two equations may be set up at each joint, only those joints with at most two unknown member forces may be considered. If the truss is statically determinate, then at least one such joint will exist at each stage of the analysis. To computerize the process of analysis, a set of general expressions for unknown forces meeting at a joint has been developed thus:

Suppose n unknown forces F_1, F_2, ..., F_n meet at a point each making an anticlockwise angle of α_1, α_2, ... α_n, ($0 < \alpha < 360$) to the reference axis (the horizontal axis in this case) as shown:

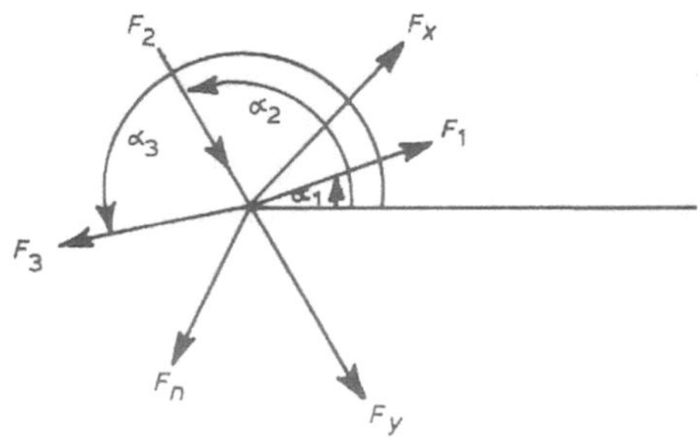

If the two unknown forces are F_x and F_y making anticlockwise angles α_x and α_y to the horizontal, then resolving horizontally,

$$\sum_{i=1}^{n} F_i \cos \alpha_i + F_x \cos \alpha_x + F_y \cos \alpha_y = 0$$

And resolving vertically,

$$\sum_{i=1}^{n} F_i \sin \alpha_i + F_x \sin \alpha_x + F_y \sin \alpha_y = 0$$

The forces F_i are assumed to be tensile positive forces, i.e. pulling on the joint, $\cos \alpha_i$ and $\sin \alpha_i$, assume the appropriate signs for the given quadrant. Solving the two equations for F_x and F_y gives the following expressions:

$$F_x = \frac{\left(\sum_{i=1}^{n} F_i \sin \alpha_i\right) \cos \alpha_y - \left(\sum_{i=1}^{n} F_i \cos \alpha_i\right) \sin \alpha_y}{\left(\cos \alpha_x \sin \alpha_y - \sin \alpha_x \cos \alpha_y\right)}$$

$$F_y = \frac{\left(\sum_{i=1}^{n} F_i \sin \alpha_i\right) \cos \alpha_x - \left(\sum_{i=1}^{n} F_i \cos \alpha_i\right) \sin \alpha_x}{\left(\cos \alpha_y \sin \alpha_x - \sin \alpha_y \cos \alpha_x\right)}$$

These expressions may be used in computer programs to carry out the method of resolution of joints.

The method of tension coefficients, on the other hand, involves assembling sufficient joint equilibrium equations and solving them simultaneously to give the force in each member. If there are N members, then N equations are required. But for a plane frame, $N = 2J - 3$; hence, there will be three superfluous equations. These can be used either to determine the reactions or to check the values of the tension coefficients obtained from the previous equations. This method avoids the problem of finding a joint with only two unknown forces as required in the joint method. It also has an advantage in the sense that the set of simultaneous equations may be solved on a computer using either the *Microsoft Excel* built-in function for solving simultaneous equations, or a computer program. The tension coefficient t of a member is the tensile force per unit length of the member (a negative value indicating compression), and the simultaneous equilibrium equations are usually written in such quantities. The equilibrium equations for member forces meeting at a joint I, shown below, can be expressed in terms of tension coefficients as follows, assuming all forces are in tension while P_x and P_y are applied joint loads. Resolving in x-direction:

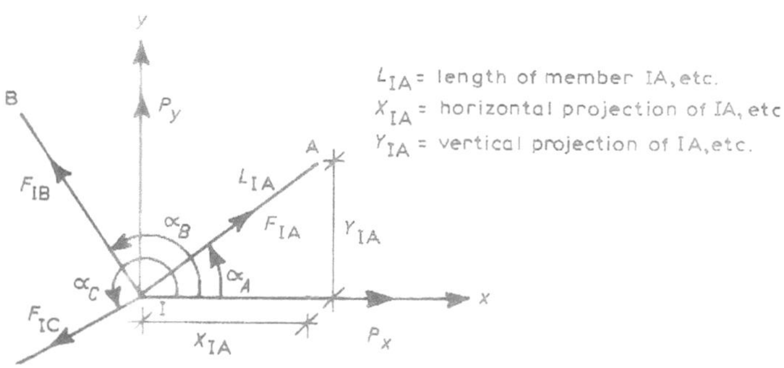

$$P_x + F_{1A}\cos\alpha_A + F_{1B}\cos\alpha_B + F_{1C}\cos\alpha_C = 0$$

$$\text{Therefore,}\quad P_x + F_{1A}\frac{X_{1A}}{L_{1A}} + F_{1B}\frac{X_{1B}}{L_{1B}} + F_{1C}\frac{X_{1C}}{L_{1C}} = 0$$

$$\text{Therefore,}\quad Px + X_{1A}t_{1A} + X_{1B}t_{1B} + X_{1C}t_{1C} = 0$$

or

$$X_{1A}t_{1A} + X_{1B}t_{1B} + X_{1C}t_{1C} = -P_x$$

This implies that the horizontal force exerted on the joint by a member is equal to its tension coefficient multiplied by the horizontal projection of the member.

Resolving in *y*-direction:

$$P_y + F_{1A}\sin\alpha_A + F_{1B}\sin\alpha_B + F_{1C}\sin\alpha_C = 0$$

$$\text{Therefore,}\quad P_y + F_{1A}\frac{Y_{1A}}{L_{1A}} + F_{1B}\frac{Y_{1B}}{L_{1B}} + F_{1C}\frac{Y_{1C}}{L_{1C}} = 0$$

$$\text{Therefore,}\quad Y_{1A}t_{1A} + Y_{1B}t_{1B} + Y_{1C}t_{1C} = -P_y$$

This implies that the vertical force exerted on the joint by a member is equal to its tension coefficient multiplied by the vertical projection of the member. It should be noted that the positive directions of the coordinate axes are arbitrary but must be the same for each joint.

In summary, the procedure in using the tension coefficient method is as follows:

i. Take positive directions for x and y.
ii. Assume all members in tension.
iii. Write down equations for each joint in the truss, observing the positive and negative sign rule depending on how the tensile forces tend to move the joints in either positive or negative directions.
iv. Solve the equations for the tension coefficients, t_{IA}, etc.
v. Check values for t_{IA} from equations of static equilibrium.
vi. Calculate the force in each truss member by multiplying the tension coefficient by the length of the member.

12.8 Truss Design

Having determined the forces on each truss member, truss design involves calculating the stresses and selecting member sizes and timber grades suitable for the purpose. Connection details are an important part of the design. Determining the deflection of the truss is also a part of the design. The design load is the total uniform load to be supported by the trusses on both sides of the roof. Wooden trusses are usually designed assuming pin-jointed members, and the load is considered to be concentrated at the panel points. However, in practice, they are assembled with bolts, nails or special connectors, while steel trusses are bolted, riveted or welded. Although these rigid joints impose secondary stresses, it is seldom necessary to consider them in the design procedure. In general, the following steps are involved in designing a truss:

1. Select general layout of truss members and truss spacing.
2. Estimate external loads to be applied including self-weight of truss, purlins and roof covering, together with wind loads.
3. Determine critical (worst combinations) loading. It is usual to consider dead loads alone, and then dead and imposed loads combined.
4. Analyse the framework to find forces in all members.
5. Select material and section to produce in each member a stress value which does not exceed the permissible value. Particular care must be taken with compression members (struts) or members normally in tension but subject to stress reversal due to wind uplift. Members in the top chord of the truss must be designed for strength in both bending and compression. Members in compression only must be designed as columns. Lower chord members must be designed for safe strength in tension and also in bending if there is a ceiling load.
6. To determine the strength of a truss member as a beam, use the following equation:

$$\frac{WL}{12} = \sigma_b S \quad \text{for a nailed truss,} \quad \text{or}$$
$$\frac{WL}{8} = \sigma_b S \quad \text{for a bolted truss}$$

where

W = the uniform load perpendicular to the member (N/mm);
L = length of the member (m);
σ_b = the safe fibre stress in bending (N/mm^2);
S = the section modulus, $S = \frac{bd^2}{6}$ (mm^3);
b = the breadth of the member (mm);
d = the depth of the member (mm).

It should be noted, however, that web members are designed for axial loads.

Worked Example on Truss Analysis and Design using the Method of Resolution at Joints

The total load on the roof of a 20 m (length) × 8 m (breadth) workshop is 38 kN. The load is to be supported by a number of 3-m-high gable trusses spaced 4 m on

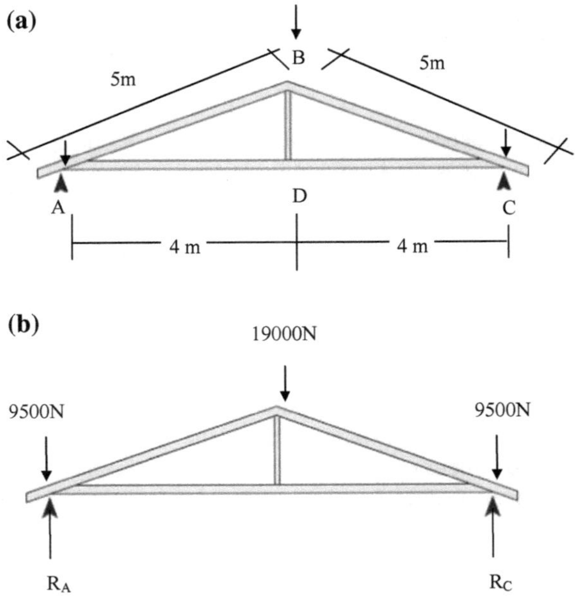

Fig. 12.10 **a** A typical gable truss for the workshop. **b** Joint load distribution on gable truss

centre and fabricated with *Melicia excelsa* of 50% grade. A typical Gable truss for the workshop is shown in Fig. 12.10a.

i. Compute the load at each joint.
ii. Determine the magnitude and type of force (i.e. tensile or compressive force) in each member of the truss using the method of resolution at joints.
iii. Determine the appropriate lumber sizes for each truss member using the Madison approach and assuming that they are short columns.
iv. If each truss is to be supported by two columns, determine the number of trusses and columns required for the workshop building.

Assume the following design values for the dry *Melicia excelsa* of 50% grade: compression parallel to grain (σ_P) = 9 N/mm^2; tension parallel to grain (σ_T) = 11.2 N/mm^2.

Solution

(i) Joint Loads

The total roof load = 38 kN
Roof load at each joint, i.e. load per joint are:

$1/2\text{W} + 1/2\text{W} + \text{W} = 38{,}000\,\text{N}$
Load at Joints A and C = 1/2 W = 9500 N
Load at Joint B = W = 19,000 N
The joint loads are indicated on the Fink truss in Fig. 12.10b.

(ii) The Magnitude and Type of Force in each Member

The magnitude and type of force (i.e. tensile or compressive force) in each member of the truss, using the method of joint resolution, are computed as follows:

Reaction forces $R_A = R_C = 38{,}000/2 = 19{,}000$ N

For convenience, it is assumed that all forces are tensile. Also, only joints with two unknown member forces are considered since only two equations of force equilibrium will be used.

Joint A = Joint C (by symmetry)

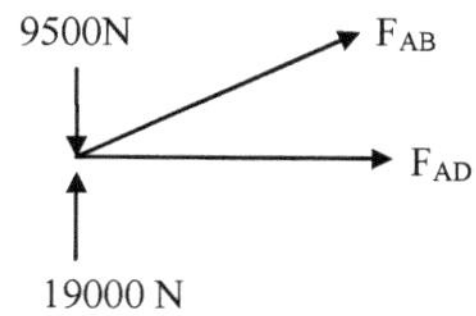

Summation of vertical forces = 0 (assuming upward forces are positive)

$R_A - 9500 + 3/5F_{AB} = 0$
$19{,}000 - 9500 + 3/5F_{AB} = 0$
$F_{AB} = F_{BC} = -15833.33\text{N} = 15.8$ kN (Compression)
Summation of horizontal forces = 0 (positive along the positive x-axis)
$F_{AD} = F_{CD}$ (by symmetry)
$4/5\ F_{AB} - F_{AD} = 0$
$F_{AD} = F_{CD} = 12666.67 = 12.7$ kN

Joint B

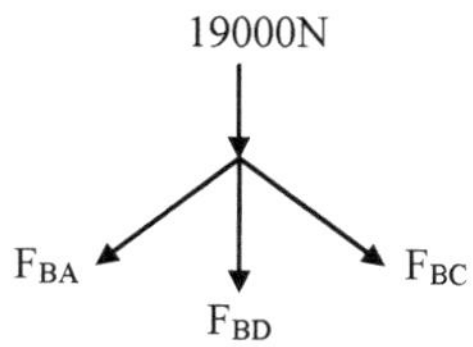

Summation of horizontal forces = 0

$4/5F_{BA} - 4/5F_{BC} = 0$
$0.8F_{BA} - 0.8F_{BC} = 0$

$F_{BA} = F_{BC} = -15833.33\ N = 15.8$ kN (compression) (this is to confirm the assumption already made

Summation of vertical forces = 0

$F_{BD} + 19{,}000 = 3/5F_{BA} + 3/5F_{BC}$
$F_{BD} = 9500 + 9500 - 19{,}000 = 0$ kN

Table 12.1 Truss member forces obtained by force resolution as joints

Member	Force (**KN**)	Type of force
AB	15.8	Compression
AD	12.7	Tension
BC	15.8	Compression
BD	0	–
CD	12.7	Tension

These results are shown in Table 12.1.

(iii) Appropriate lumber sizes for each truss member assuming short columns

For compression members AB and BC
$P/A \leq 9\ N/mm^2$
Assuming a 25 × 50 mm lumber cross-section is selected
$15{,}800/(25 \times 75) = 8.4\ N/mm^2$ OK

For tension members AD and CD
$P/A \leq 11.2\ N/mm^2$
Assuming 25 × 50 mm lumber cross-section is selected
$12{,}700/(25 \times 75) = 6.8\ N/mm^2$ OK

(iv) Number of trusses and columns required for the workshop building

$$\begin{aligned}\text{The number of trusses required} &= (\text{Length of building}/\text{truss spacing}) + 1\\ &= 20/4 + 1 = 6\,\text{Gable trusses}\\ \text{The number of columns required} &= \text{Number of trusses} \times 2\\ &= 6 \times 2 = 12\,\text{Columns}\end{aligned}$$

Worked Example 2: Using the Method of Tension Coefficients

Determine the forces in the members of the gable truss whose joint loads were earlier determined as indicated in Fig. 12.10b reproduced below, using the method of tension coefficients. Assume that the trusses will be fabricated with dry *Melicia excelsa* of 50% grade and spaced 4 m on centre in a roof system. Determine the appropriate lumber sizes for each truss member using the Madison approach and assuming that they are short columns. Assume the following design values for the dry *Melicia excelsa* of 50% grade: compression parallel to grain (σ_P) = 9 N/mm^2; tension parallel to grain (σ_T) = 11.2 N/mm^2.

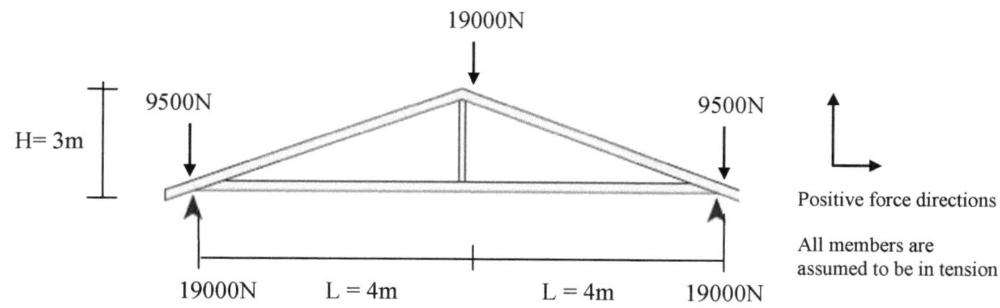

Solution

We need a total of 5 equilibrium equations (containing the 5 unknown forces) which are to be solved simultaneously. The positive directions of forces are shown in the figure above, and these axes may be assumed to be attached to the particular joint under consideration. As we move from joint to joint, these directions would not be changed. All members are assumed to be in tension, and the direction of action of these forces will either be positive or negative according to the chosen force direction axes. Thus, assembling the equations of equilibrium using tension coefficients:

Note that at each joint, summation of vertical forces = 0 (assuming upward forces are positive); summation of horizontal forces = 0 (positive along the positive *x*-axis); horizontal and vertical projections of the forces are 4 and 3 m, respectively.

Joint A

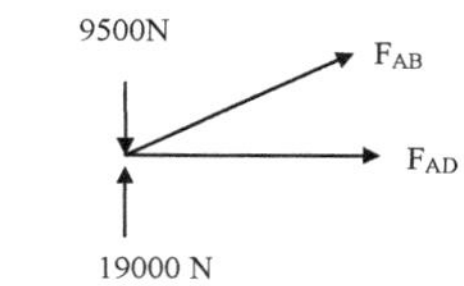

$19000 - 9500 + 3t_{AB} = 0$

$$\text{i.e., } 3t_{AB} = -9500 \quad \text{(i)}$$

$$4t_{AB} + 4t_{AD} = 0 \quad \text{(ii)}$$

Joint B

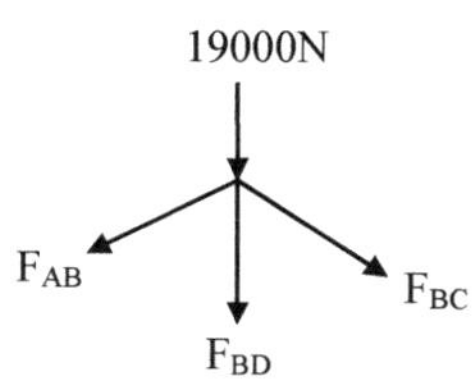

$$1900 + 3t_{AB} + 3t_{BC} + 3t_{BD} = 0 \qquad \text{(iii)}$$

$$-4t_{AB} + 4t_{BC} = 0 \qquad \text{(iv)}$$

Joint C

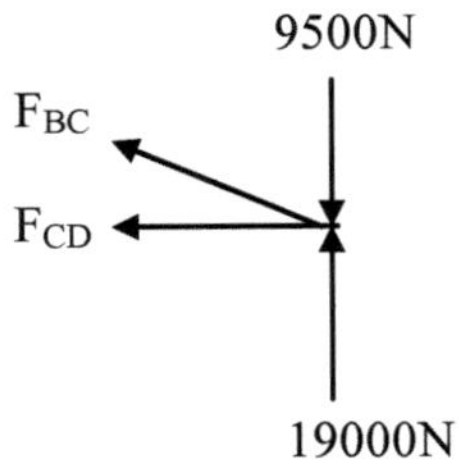

$$3t_{BC} + 19,000 - 9500 = 0 \qquad \text{(v)}$$

$$4t_{CD} + 4t_{BC} = 0 \qquad \text{(vi)}$$

Collecting the five of the six equations (by inspection) to be able to determine the five member forces, we have:

$$3t_{AB} = -9500 \qquad \text{(i)}$$

$$4t_{AB} + 4t_{AD} = 0 \qquad \text{(ii)}$$

$$1900 + 3t_{AB} + 3t_{BC} + 3t_{BD} = 0 \qquad \text{(iii)}$$

$$-4t_{AB} + 4t_{BC} = 0 \qquad \text{(iv)}$$

$$4t_{CD} + 4t_{BC} = 0 \qquad \text{(v)}$$

For easy comprehension, if we let $t_{AB} = X_1$, $t_{AD} = X_2$, $t_{BC} = X_3$, $t_{BD} = X_4$, and $t_{CD} = X_5$

Then, the five equations can be rewritten as follows:

$$3X_1 = -9500 \qquad \text{(i)}$$

$$4X_1 + 4X_2 = 0 \qquad \text{(ii)}$$

$$3X_1 + 3X_3 + 3X_4 = -1900 \qquad \text{(iii)}$$

$$-4X_1 + 4X_3 = 0 \qquad \text{(iv)}$$

$$4X_3 + 4X_5 = 0 \qquad \text{(v)}$$

These simultaneous equations in the form Ax = B can be converted into a 5 × 5 square matrix and solved using the in-built *inverse* and *matrix multiplication* functions of *Microsoft Excel*. In doing so, the following procedure should be followed:

i. Note that the matrix coefficients are the vertical and horizontal projections of the truss members at each joint; the B vector is the RHS value at each joint; the X vector are the computed values of the tension coefficients; while the product of the tension coefficient and the actual length of each member is the force in the member.

ii. **To find the inverse of a matrix**

- 1. Pre-select and highlight the cells, i.e. the rows and columns for the inverse matrix.
- 2. Enter in the formula bar the function (= **MINVERSE**(array of cell references)), e.g. = **MINVERSE**(A4:H11).
- 3. Press **Control + Shift + Enter** together.

iii. **To multiply two matrices**

- Pre-select and highlight the cells, i.e. the rows and columns for the matrix product.
- Enter in the formula bar the function (= **MMULT**(cell references)), e.g. **MMULT**(J4:Q11,S4:S11).
- Press **Control + Shift + Enter** together.

The Excel spreadsheet result is shown in Fig. 12.11a, b. The values tally with those obtained with the method of joint resolution.

12.9 Storage Handling and Bracing of Wooden Trusses

Trusses are slender elements. They are very strong when placed in the vertical position, but can be easily damaged or broken if racked or bent in the lateral direction. Damage or failure can occur at the joints or within the lumber members. When wooden trusses arrive at the job site, they should be checked for any permanent damage such as cross breaks in the lumber, missing or damaged metal connector plates, excessive splits in the lumber or any damage that could impair their structural integrity.

(a)

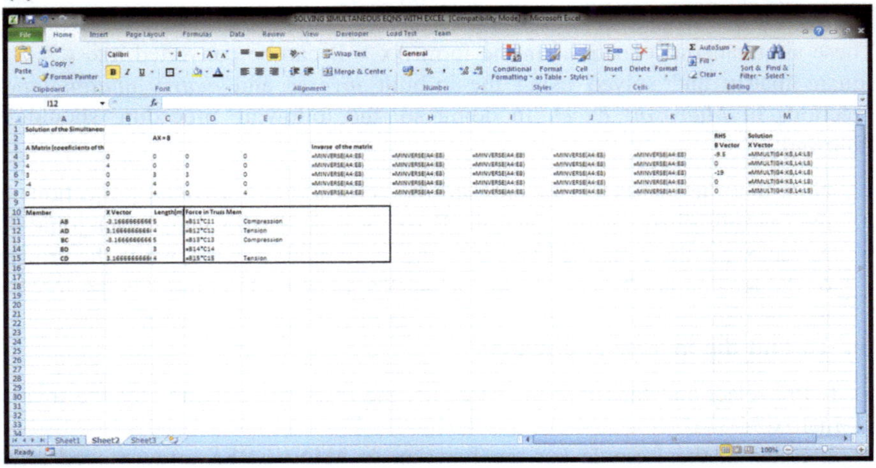

(b)

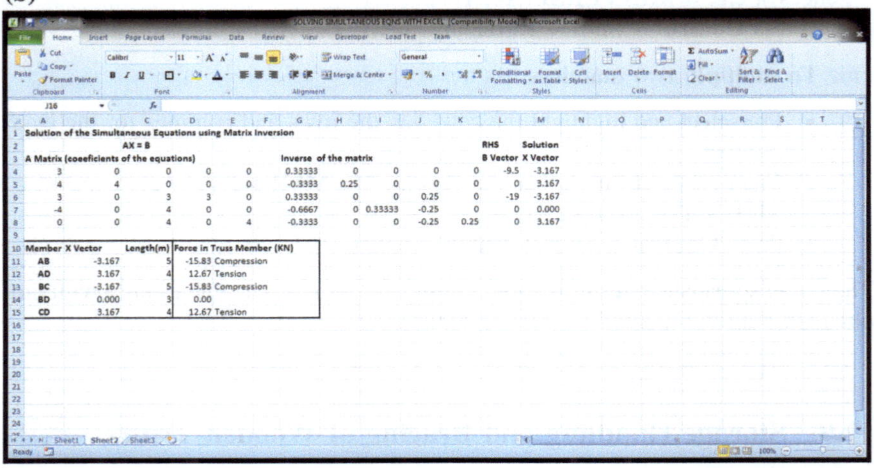

Fig. 12.11 **a** *Excel* worksheet showing the formulas for the computations. **b** *Excel* worksheet showing the values of the member forces obtained

12.10 Truss Support

Temporary and permanent bracings are essential components that require design and consideration. Bearing walls and lintels must be designed to resist the truss loads, including point loads from girders. The building designer should also ensure that all nails, hangers and uplift anchors between trusses and support framing are adequate. Typically, prescriptive toe nailing requirements are given in building

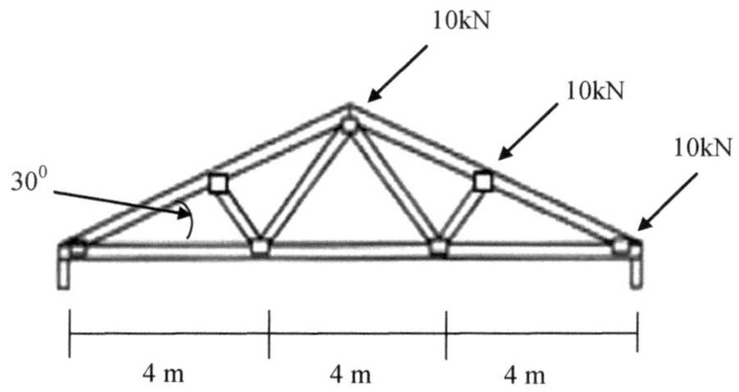

Fig. 12.12 A loaded fink truss

codes for smaller structures, whereas metal uplift anchors are often used for larger buildings or in areas of high wind.

12.11 Preservative Treatment and Truss Plates

Special attention should be given to service conditions such as wet service and an associated treated service condition. Truss plates should not be used in incised lumber. Agricultural buildings can sometimes give rise to corrosive environments that require special measures such as a higher standard of galvanization.

Practice Questions

1. A Fink truss is loaded as shown in Fig. 12.12. Determine the force in the truss members.
2. Compute the magnitude and type (i.e. tension or compression) forces in the members of the Howe truss of Worked Example 1 reproduced below. Determine the appropriate lumber sizes for each truss member assuming that they are short columns. Also, assume the following design values: compression parallel to grain (σ_P) = 9 N/mm^2; tension parallel to grain (σ_T) = 11.2 N/mm^2.

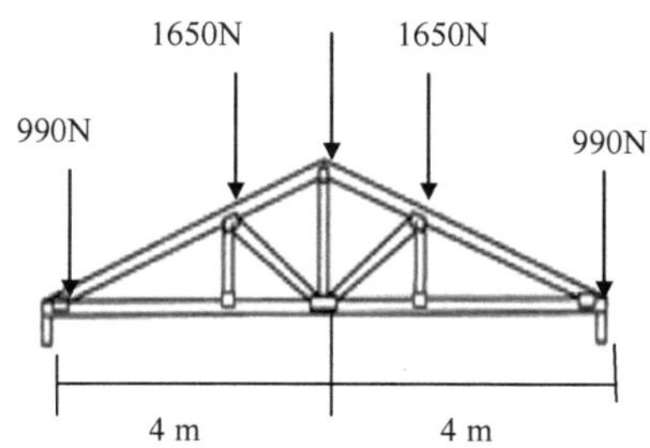

3. Write a computer code for computing the values of two unknown forces meeting at a joint.

References

Andrews KRF (1986) Basic theory of structures. Butterworth & Co Publishers, London

Breyer DE, Fridley KJ, Cobeen KE (1999) Design of wood structures, 4th edn. McGraw-Hill, New York, USA

Faherty KF, Williamson TG (1999) Wood engineering and construction handbook, 3rd edn. McGraw-Hill Handbooks, New York, USA

Goldstein EW (1999) Timber construction for architects and builders. McGraw-Hill Publishers, New York, USA

Gurfinkel G (1973) Wood engineering, published by Southern Forest Products Association, USA

Khurmi RS, Khurmi N (2013) Strength of materials (Mechanics of solids in S.I. units, international edition, S.Chand & Company PVT Ltd., New Delhi, India

Lindley JA, Whitaker JH (1996) Agricultural buildings and structures, revised edition. American Society of Agricultural Engineers, USA

Lucas E, Olorunnisola A, Adewole N (2006) Preliminary evaluation of guava (Psidium guajava L.) tree branches for truss fabrication in Nigeria. Agricultural Engineering International: The CIGR E-journal. Manuscript No. BC 05 010, 8:1–10

Marshall WT, Nelson HM (1978) Structures. Pitman Publishing Ltd., London

Midwest Plan Service (1977) Structures and environment handbook. Midwest Plan Service, Ames, Iowa, USA

NCP (2005) Nigerian code of practice for structural timber design. Standards Organisation of Nigeria

Ozelton EC, Baird JA (1976) Timber designers' manual. Granada Publishing Limited, London

Seward D (2003a) Understanding structures, analysis, materials, design, 3rd edn. Palgrave Macmillan, New York, USA

Seward D (2003b) Understanding structures, analysis, materials, design, 3rd edn. Palgrave Macmillan, New York, USA

Shaeffer RE (1999) Elementary structures for architects and builders, 4th edn. Prentice-Hall Publishing Co., New Jersey, USA

Chapter 13
An Introduction to Wooden Bridges

A bridge can be simply defined as a structure built to span a gorge, valley, road, railroad track, river, body of water or any other physical obstacle for the purpose of providing passage over the obstacle. A wooden bridge is one that uses wood as its principal structural material. The most ancient man-made wooden bridges were most probably log bridges, created by letting trees fall over the river. Wooden bridges continue to be relevant to date because:

- Wood is readily available in many parts of the world. Also, the materials for timber bridges are readily available and relatively cheap in most cases. They can be found at the site or purchased locally.
- The strength of timber, its light weight and energy-absorbing properties make it highly desirable for bridge construction.
- Timber can meet the same structural requirements as any other construction material including conformance with standard codes of practice.
- Wood is inherently durable when properly protected against rotting, shrinking, twisting, insect attack and everyday exposure to weather elements. Wood can now be effectively protected for periods of 50 years or longer. Properly treated wood will not crumble like concrete, will not rust like steel, and can be used in any environment regardless of climate. Again, wood is not damaged by continuous freezing and thawing and can resist harmful effects of de-icing agents which cause deterioration in other bridge materials.
- Contrary to popular belief, large wood members provide good fire resistance qualities that meet or exceed those of other materials in severe fire exposures.
- Wooden bridges are environmentally friendly compared to other bridge types.
- Timber bridge construction is usually quick. Little site preparation is needed and the construction can occur in virtually any weather condition without detriment to the material.
- Wooden bridges are relatively economical because wood is competitive with other structural materials on a first cost basis and shows advantages when life cycle costs are compared.

A.O. Olorunnisola, *Design of Structural Elements with Tropical Hardwoods*,
DOI 10.1007/978-3-319-65343-3_13

- Wooden bridges can be placed over small streams or channels with firm, stable banks.
- Using fallen trees, stringer bridges can be built. New lumber and wood scavenged from buildings and rail road ties can also be used to build stringer bridges.
- Technological advances in laminating over the past five decades have further increased the suitability and performance of wood for modern vehicular bridges. Glulam in particular provides greater strength than sawn wood for longer span application.
- Wooden bridges require little maintenance.

Some of the disadvantages of wooden bridges include the fact that they may pose safety hazards if not properly designed and/or constructed, logs can have rots, knots and other problems that affect strength, and the use of untreated lumber can create durability challenges.

Log bridges are still common in many rural areas of Africa today where they usually carry pedestrians on single logs (Fig. 13.1a) and even vehicles on multiple parallel logs (Fig. 13.1b) that either fell naturally or were intentionally fell and placed across streams. The span of such brides is usually short. The use of emplaced logs is also sometimes employed in temporary bridges used for logging roads, where a forest tract is to be harvested and the road then abandoned. Such log bridges have a severely limited lifetime due to soil contact and subsequent rot and insect infestation. Materials of construction of relatively more durable log bridges include logs and dry set stonework footings. The top may be flattened or boards added, topped with rammed earth for vehicles. Even longer lasting log bridges may be constructed by using treated logs and/or by providing well drained footings of stone or concrete combined with regular maintenance to prevent soil infiltration.

Another common type of wooden bridge is the beam bridge (shown in Fig. 13.2a, b). This is the simplest structural form for bridge spans supported by an abutment or pier at each end. It is regarded as a simply supported bridge since no moments are transferred throughout the support.

The simplest beam bridge could be a wood plank laid across a stream. Types of construction could include having many beams side by side with a deck across the top of them, or a main beam on either side supporting a deck between them. The main beams could be I-beams (or H-beams), trusses, or box girders. They could be half-through or braced across the top to create a through bridge. Beam bridges are not limited to a single span. Some have multiple simply supported spans supported by piers. Such bridges are often only used for relatively short distances because, unlike truss bridges, they have no built-in supports. The only supports are provided by piers. The farther apart its supports, the weaker a beam bridge gets. As a result, beam bridges rarely span more than 75 m (250 feet). This does not mean they cannot be used to cross great distances, it only means that a series of beam bridges must be joined together to create, a continuous span.

Another type of wooden bridge is a covered bridge with a roof and siding which, in most cases, create an almost complete enclosure (Fig. 13.3). The purpose of the covering is to protect the wooden members from the weather. Uncovered wooden

(a)

(b)

Fig. 13.1 **a** A single log bridge. **b** A multiple-log bridge

(a)

(b)

Fig. 13.2 **a** A shot-span wooden beam bridge. **b** A long-spanwooden beam bridge

bridges have a lifespan of only 10–15 years because of the effects of rain and sun. However, bridges covered for other reasons such as protecting pedestrians and keeping horses from shying away from water, are also sometimes called covered bridges.

Early covered bridges consisted of horizontal beams laid on top of piles driven into the riverbed. The challenge with such bridges is that the length between spans is usually limited by the maximum length of each beam which is about 9 m (30

Fig. 13.3 A covered wooden bridge

feet). The development of timber trusses allowed bridges to span greater distances than those with timber beam only.

A truss bridge (Fig. 13.4a) is one whose load-bearing superstructure is composed of a truss. Truss bridges represent one of the oldest types of modern wooden bridges. The basic types have simple designs which can be easily analysed. The ability to distribute the forces in various ways has led to a large variety of truss bridge types. Some types may be more advantageous when wood is employed for compression elements while other types may be easier to erect in particular site conditions, or when the balance between labour, machinery and material costs have certain favourable proportions. In general, wooden truss bridges are economical to construct because of the efficient use of materials. Early versions used king post and queen post configurations, while some included diagonal panel bracing with parallel top and bottom chords. Another design configuration is one that has carefully fitted timbers as compression members and iron rods as tension members (Fig. 13.4b), with the general features of a covered bridge to protect the structure. A truss bridge may carry its roadbed on top, in the middle, or at the bottom. When the roadbed is atop the truss it is called a deck truss. When the truss members are both above and below the roadbed, it is called a through truss, and where the sides extend above the roadbed but are not connected, a half-through truss. When the roadbed is atop the truss it is called a deck truss. Bridges with the roadbed at the top or the bottom are the most common as this allows both the top and bottom to be stiffened, forming a box truss. In general, wooden truss bridges could be up to 12 m (40 feet) long with cross-sections up to 100 × 300 mm (4 × 12 inches). In most cases, their design, fabrication and erection are relatively simple. However, once assembled, they take up a relatively great amount space.

(a)

(b)

Fig. 13.4 **a** A typical wooden truss bridge. **b** A wooden truss bridge with iron rods as tension members

References

Faherty KF, Williamson TG (1999) Wood engineering and construction handbook, 3rd edn. McGraw-Hill Handbooks, New York, USA

Goldstein EW (1999) Timber construction for architects and builders. McGraw-Hill Publishers, New York, USA

Gurfinkel G (1973) Wood engineering, published by Southern Forest Products Association, USA

Ozelton EC, Baird JA (1976) Timber designers' manual. Granada Publishing Limited, London

Appendix A
A Brief Review of Simple Microsoft Excel Operations

Excel is a software program designed to help the user evaluate and present information in a spreadsheet format. Spreadsheets are most often used by business for cash flow analysis, financial reports and inventory management. Before the era of computers, a spreadsheet was simply a piece of paper with a grid of rows and columns to facilitate entering and displaying information. Excel is so flexible that its application can extend beyond traditional spreadsheets into the area of data analysis. You can use Excel to enter data, analyse the data with basic statistical tests and charts, and then create reports.

This brief review introduces the reader to how to work with Excel 2010 in the Windows operating system. You will be introduced to basic workbook concepts, including navigating through your worksheets and worksheet cells. This book requires prior Excel experience. Familiarity with basic features of that program will reduce your start-up time. This section provides a quick overview of the features of Excel.

Excel Workbooks and Worksheets

Excel documents are called **workbooks**. Each workbook is made up of individual spreadsheets called **worksheets**. A single workbook can have as many as 255 worksheets. The names of the sheets appear on tabs at the bottom of the workbook window.

Launching Excel

With Excel 2010 installed on your computer, the installation program automatically inserts a shortcut icon to Excel 2010 in the Programs menu located under the Windows Start button. You can click this icon to launch Excel.

To start Excel

- Click the Start button on the Windows Taskbar and then click '***All Programs***'.
- Click ***Microsoft Office*** and then click ***Microsoft Office Excel 2010***.
- Excel starts up, displaying window shown in Fig. A.1. Your window might look different depending on how Excel has been set up on your system. Before proceeding, take time to review the various elements of the Excel window.

A.O. Olorunnisola, *Design of Structural Elements with Tropical Hardwoods*,
DOI 10.1007/978-3-319-65343-3

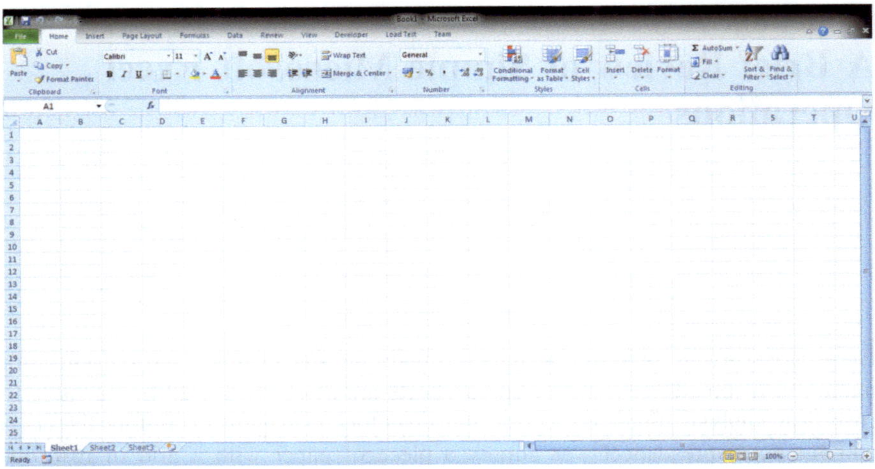

Fig. A.1 An excel spreadsheet window

Viewing the Excel Window

The Excel window shown in Fig. A.1 is the environment in which you may carry out structural analysis and/or design. Your window might look different depending on how Excel has been set up on your system. Before proceeding, take time to review the various elements of the Excel window.

Running Excel Commands

You can run an Excel command either by clicking the icons found on the Excel ribbon or by clicking the Office button and then clicking one of the commands from the menu that appears. Note that some of the commands have keyboard shortcut key combinations that run a command or macro. For example, pressing the CTRL and following keys simultaneously will also run the following commands:

CTRL + A = Select all items in a worksheet
CTRL + B = Bolden selected items in a worksheet
CTRL + C = Copy selected items in a worksheet
CTRL + F = Find a particular item in a worksheet
CTRL + G = Go to a particular location/item in a worksheet
CTRL + H = Find and replace a particular item in a worksheet
CTRL + I = Convert selected items in a worksheet into italics
CTRL + K = Insert a hyperlink in a worksheet
CTRL + L = Create a table in a worksheet
CTRL + N = Create a new blank document
CTRL + O = Open a new blank document
CTRL + P = Print a document
CTRL + S = Save a document

CTRL + T = Create a table in a worksheet
CTRL + U = Underline selected items in a worksheet
CTRL + V = Paste selected items in a worksheet
CTRL + X = Cut/delete selected items in a worksheet
CTRL + Z = Undo a previous action
CTRL + Enter = Replicate text values or formulas over cell ranges
CTRL + PgUp/PgDn= Move to next/previous worksheet
CTRL + F12 = Open existing worksheet.

The menu commands below the Office button are used to set the properties of your Excel application and entire Excel documents. If you want to work with the contents of a document, you work with the commands found on the Excel ribbon. Each of the tabs on the Excel ribbon contains a rich collection of icons and buttons providing one-click access to Excel commands.

Scrolling through a Workbook

To move from one sheet to another, you can either click the various sheet tabs in the workbook or use the navigational buttons located at the bottom of the workbook window.

Worksheet Cells

Each worksheet can be thought of as a grid of cells, where each cell can contain a numeric or text entry. Cells are referenced by their location on the grid, i.e. column label (A, B, C, D and so on) and row number (1, 2, 3 and so on). An example of a cell reference is A1, i.e. a cell located in Column A, Row 1. As you will see later, if you were to use this value in a function or Excel command, you would use the cell reference. An active Excel cell address usually appears in the Name box

Selecting a Cell

When you want to enter data or format a particular value, you must first select the cell containing the data or value. To do this, you simply click on the cell in the worksheet. If you want to select a group of cells, known as a cell range or range, you must select one corner of the range and then drag the mouse pointer over cells. To see how this works in practice, try selecting figures pre-written and located in the cell range A1:C17. To select a cell range:

- Click A1
- With the mouse button still pressed, drag the mouse pointer over to cell C17
- Release the mouse button.

Now the range of cells from A1 down to C17 is selected. Observe that a selected cell range is highlighted to differentiate it from unselected cells. A cell range selected in this fashion is always rectangular in shape and contiguous. If you want to select a range that is not rectangular or contiguous, you must use the CTRL key on your keyboard and then select the separate distinct groups that make up the range. For example, if you want to select only the cells in the range B4:B17 and F4: F17, you must use this technique:

- Select the range B4:B17
- Press the CTRL key on your keyboard
- With the CTRL key still pressed, select the range F4:F17.

Data Entry

To enter data into an Excel worksheet, you first select the cell corresponding to the upper left corner of the table, making it the active cell. You then type the value or text you want to place in the cell. To move between cells, you can either press the Tab key to move to the next column in the same row or press the Enter key to move to the next row in the same column. If you are entering data into several columns, the Enter key will move you to the next row in the first column of the data set.

Data formats are the fonts and styles that Excel applies to your data's appearance. Formats are applied to either text or numbers. You can access all of the possible formatting options for a particular cell by opening the Format Cells dialogue box.

Moving Cells

Excel allows you to move the contents of your cells around without affecting their values. This is a great help in formatting your worksheets. To move a cell or range of cells, simply select the cells and then drag the selection to a new location. You can also use the Cut, Copy and Paste buttons to move a cell range. These buttons are essential if you want to move a cell range to a new workbook or worksheet.

Formulas and Functions

Not all of the values displayed in a workbook come from data entry. Some values are calculated using formulas and functions. A formula always begins with an equal sign followed by a function name, number, text string or cell reference. Most functions contain mathematical operators such as 1or 2. A list of mathematical operators is shown below:

Operator	**Description**
+	Addition
−	Subtraction
/	Division
*	Multiplication
^	Exponentiation.

Inserting a Simple Formula

To see how to enter a simple formula, type 23 in cell A1 and press Enter and type 23 in cell B1 and press Enter. To add both figures, Type =A1+B1 in cell C1 and

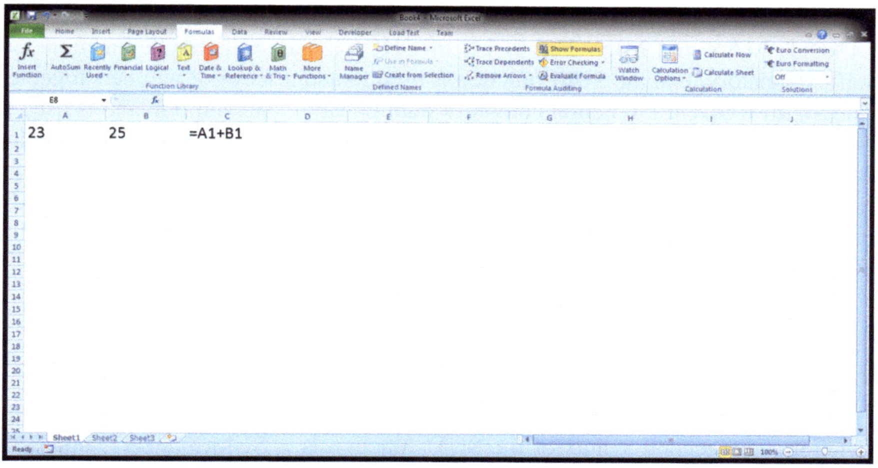

Fig. A.2 Inserting formula in excel

press Enter. The formula is displayed in C1. The sum of these two values is shown in Fig. A.3 (see Fig. A.2).

Inserting an Excel Function

Excel has a library containing hundreds of functions covering most mathematical needs. Users can also create their own custom functions using Excel's programming language. A function is composed of the function name and a list of arguments, i.e. values required by the function. The common Excel functions required for

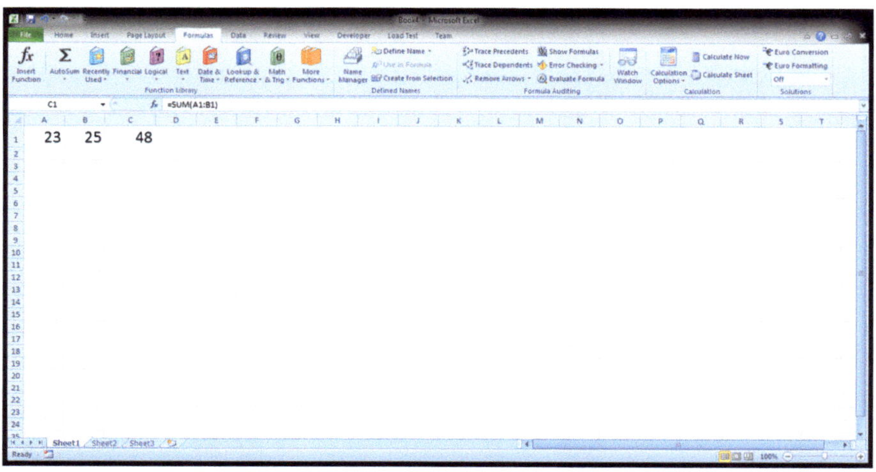

Fig. A.3 Addition result in excel

structural analysis include SUM, PRODUCT, SUMPRODUCT, AVERAGE, COUNT, MAX and MIN. The SUM, PRODUCT and AVERAGE functions are self-explanatory, e.g. the SUM function is used to add a series of numbers together. The SUMPRODUCT function combines multiplication and addition and is used to add two columns of figures that are first multiplied together. The COUNT function is useful when dealing with a very large cell range, and you need to know how many items it contains. The MAX function returns the highest value, while the MIN function returns the lowest value in a list of arguments (i.e. the text within parentheses as will be shown shortly).

The general form or syntax of an Excel function is

$$\text{FUNCTION}(\text{number1}, \text{number2}, \ldots)$$

For example, to calculate the sum of a set of cells, you would use the SUM function. The general syntax of the SUM function is = SUM(number1, number2, …) where number1 and number2 are the arguments and may be cell references [e.g. (A1, B1)], values (e.g. 23, 25) or even a simple calculation (e.g. 50*B1). For example, to calculate the sum of the cells in the range A2:B1, you could enter the formula = SUM(A1:B1) as shown in Fig. A.4.

Note that the SUM function allows multiple numbers of cell references, i.e. the number1 and number2 arguments may refer to ranges of cells that contain numbers or formulas. For example, if you wish to compute the grand total of the values in three different columns of a worksheet, say in the ranges A1:B5, C1:D5 and E1:F5 with the SUM function, all that you need to do is to replace the number1 argument with the first cell range, A1:B5; the number2 argument with the second cell range,

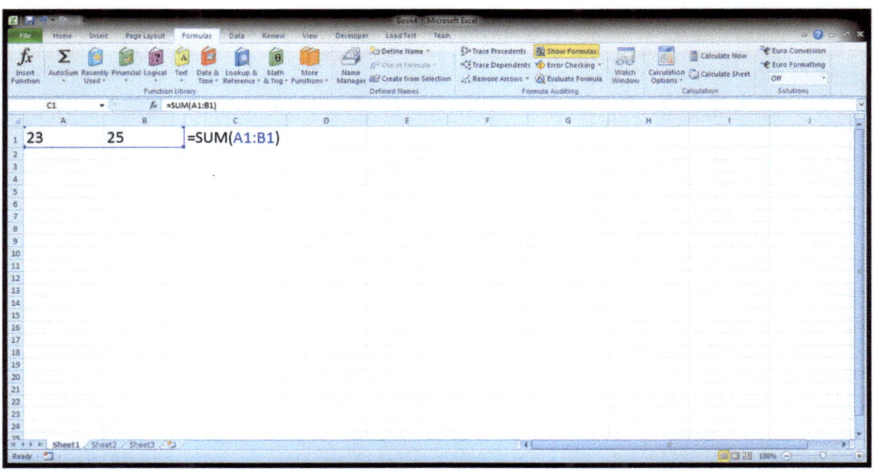

Fig. A.4 Using an excel function

C1:D5, and the number3 argument with the third cell range, E1:F5. Then enter the following formula in the cell where you want the total of the numbers in these cells to appear:

$$= \text{SUM(A1:B5, C1:D5, E1:F5)}$$

When creating formulas like these, you must separate each argument with a comma but here should be no space in-between. Although you can type in functions directly, you may find it easier to use the commands located in the Function Library group on the Formulas tab. These commands provide information on the parameters required for calculating the function value as well as giving one-click access to online help regarding each function.

Cell References

A **relative reference** identifies a cell range on the basis of its position relative to the cell containing the formula. One advantage of using relative reference is that you can fill up a row or column with a formula, and the cell references in the new formulas will shift along with the cell. If you do not want Excel to shift the cell reference when you copy the formula into other cells; if you want the formula always to point to a specific cell in your worksheet, you will need an **absolute reference**. In an absolute reference, the cell reference is prefixed with dollar signs. For example, the formula SUM($C$2:$C$11) is an absolute reference to the range C2:C11. If you copied this formula into other cells, it would still point to C2:C11 and would not be shifted. You can also create formulas that use mixed references, combining both absolute and relative references. For example, the formulas SUM ($C2:$C11) and SUM(C$2:C$11) use mixed references. In the first example, the column is absolute but the row is relative, and in the second example, the column is relative but the row is absolute. This means that in the first example, Excel will shift the row references but not the column references, and in the second example, Excel will shift the column references but not the row references. You can learn more about reference types and how to use them in Excel's online Help.

Range Names

Another way of referencing a cell in your workbook is with a range name. Range names are names given to specific cells or cells ranges. For example, you can define the range name Weight to refer to cells B2:B11 in your worksheet. To calculate the total weight, you could use the formula SUM(B2:B11) or SUM(Weight). Range names have the advantage of making your formulas easier to write and interpret. Without range names, you would have to know something about the worksheet before you could determine what the formula = SUM(B2:B11) calculates. Excel provides several tools to create range names. A simple way to create range names is to select the range of data including a row or column of titles. You can then use the titles from the worksheet to define the range name.

Sorting Data

Once you have entered your data into Excel, you are ready for computation. One of the simplest analyses is to determine the range of the data values. Which values are largest? Which are smallest? To answer questions of this type, you can use Excel to sort the data. For example, you can sort the weight data in descending order, displaying first the item that has the greatest weight down through the item that has had the smallest weight.

Printing from Excel

Before sending a job to the printer, it is usually a good idea to preview the output. With Excel's Print Preview window, you can view your job before it is printed, as well as set up the page margins, orientation, and headers and footers. To preview a print job: click the Office button, then click Print, and then click Print Preview. Click the Close Print Preview button from the Preview group on the Print Preview tab to close the Preview window.

To print your worksheet, you can select the Print command from the Office menu or Press CTRL + P. The Print dialogue box appears as shown in Fig. A.5. Notice that you can print a selection of the active worksheet, the entire active sheet or sheets, or the entire workbook. You can also select the number of copies to print and the range of pages. The other options let you select your printer from a list (if you have access to more than one) and set the properties for that particular printer. You can also click the Preview button to go to the Print Preview window. Click OK to start the print job.

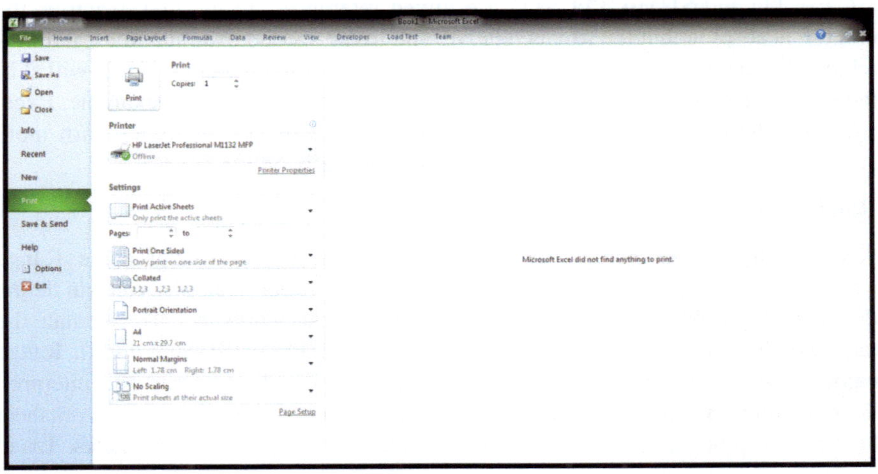

Fig. A.5 Print dialogue box

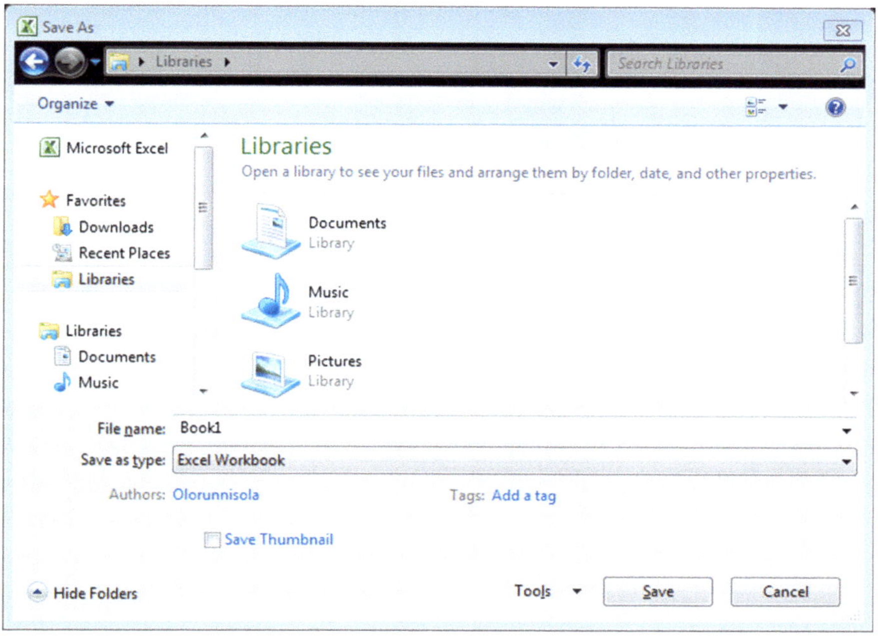

Fig. A.6 Typical 'Save As' dialogue box

Saving Your Work

You should periodically save your work so as not to lose much work if your computer or Excel crashes. Excel offers two options for saving your work: the Save command, which saves the file; and the Save As command, which allows you to save the file under a new name. A Save As dialogue box is shown in Fig. A.6.

Exiting Excel

To exit Excel: click the Office button and then click the Exit Excel button from the bottom of the menu. If you have unsaved work, Excel asks whether you want to save it before exiting. If you click No, Excel closes and you lose your work. If you click Yes, Excel opens the Save As dialogue box and allows you to save your work. Once you have closed Excel, you are returned to the Windows desktop or to another active application.

Appendix B

Strength groupings of selected Nigerian-grown timber species

N_1	N_2	N_3	N_4	N_5	N_6	N_7
Afrormosia elata	*Afzelia Africana*	*Isoberlinia doka*	Daniellia ogea	*Daniellia oliveri*	Acacia albida	Alstonia boonei
Celtis zenkeri	Nesogordonia papaverifera	*Khaya gradifoliola*	Entandrophragma angolense	Gmelina arborea	Antiaris africana	
Combretodendron macrocarpum	Erythrophleum ivorense	*Khaya senegalensis*	Lovoa trichiloides	*Khaya ivorensis*	Cordia platythyrsa	
Lophira alata	Sterculia rhinopetala	*Melicia excelsa*	Pterygota macrocarpa	Gossweilerodendron balsamiferum	Canarium schweinfurthii	
Nauclea diderrichii	Strombosia pustulata	*Tectona grandis*	Terminalia *ivorensis*		Terminalia superba	
		Guarea cedrata	Mitragyna ciliata		*Triplochyton scleroxylon*	
		Mansonia altissima				
		Holoptelia grandis				

A.O. Olorunnisola, *Design of Structural Elements with Tropical Hardwoods*,
DOI 10.1007/978-3-319-65343-3

Index

A.O. Olorunnisola, *Design of Structural Elements with Tropical Hardwoods*,
DOI 10.1007/978-3-319-65343-3

T

U

V

W

Z

MIX
Papier aus verantwortungsvollen Quellen
Paper from responsible sources
FSC® C105338

If you have any concerns about our products,
you can contact us on
ProductSafety@springernature.com

In case Publisher is established outside the EU,
the EU authorized representative is:
Springer Nature Customer Service Center GmbH
Europaplatz 3, 69115 Heidelberg, Germany

Printed by Libri Plureos GmbH
in Hamburg, Germany